HOW SAFE IS SAFE ENOUGH?

HOW SAFE IS SAFE ENOUGH?

Technological Risks, Real and Perceived

E. E. Lewis

ARREL

BOOKS

Copyright © 2014 by Elmer E. Lewis

All Rights Reserved. No part of this book may be reproduced in any manner without the express written consent of the publisher, except in the case of brief excerpts in critical reviews or articles. All inquiries should be addressed to Carrel Books, 307 West 36th Street, 11th Floor, New York, NY 10018.

Carrel Books may be purchased in bulk at special discounts for sales promotion, corporate gifts, fund-raising, or educational purposes. Special editions can also be created to specifications. For details, contact the Special Sales Department, Carrel Books, 307 West 36th Street, 11th Floor, New York, NY 10018 or carrelbooks@skyhorsepublishing.com.

Carrel Books® is a registered trademark of Skyhorse Publishing, Inc.®, a Delaware corporation.

Visit our website at www.carrelbooks.com

10 9 8 7 6 5 4 3 2 1

Library of Congress Cataloging-in-Publication Data is available on file.

Print ISBN: 978-1-63144-001-4
Ebook ISBN: 978-1-63144-016-8

Printed in the United States of America

CONTENTS

To Ann

CHAPTER 1

Technological Risk—the Past as Prologue

A RUNAWAY TRAIN turns a Canadian town into an inferno. Unexplained battery fires ground America's most advanced airliner. A Gulf oil rig explosion creates environmental havoc. Radioactivity leaks from reactors in Japan. No wonder we question our pervasive dependence on technology and fret about undiscovered hazards that may lurk behind its continued development. As technology becomes more complex, it seems that the potential for disaster looms larger, and threats to our health and the environment grow more insidious. Contemplating such threats often brings about a yearning for earlier times and a tendency to romanticize a life that was freer from the risks that technology has wrought. Before industrialization, the technology that existed was much simpler and easier to understand, such as wagons, waterwheels, and windmills. We may long for freedom from the accidents and contamination that arise from the less comprehensible complexities of today's technology, and with that yearning come feelings that we have pushed too fast and too far.

Such yearnings for a more agrarian past aren't unique to the twenty-first century, but go back at least to the early days of the Industrial Revolution. Nostalgia for an idyllic life in a pre-industrial Eden has been a persevering theme in literary circles since the poetry of Wordsworth and Blake and can be traced through the years to critics of today's technology. But with closer examination, the allure of earlier, less technological times becomes clouded. Arguably, Renaissance artist Albrecht Dürer symbolized those times most perceptively in his woodcarving *Four Horsemen of the Apocalypse*. The four—Death and its three pervasive causes, Famine, Pestilence, and War—shadowed the lives of our forebears. Theirs were lives in which half of the children born wouldn't live to reach adulthood, and where few adults could expect to reach what today we consider old or even middle-aged. Life was precarious, often suddenly and unexpectedly cut off in its prime by famine, pestilence, or war.

Figure 1.1: Albrecht Durer, *The Four Horsemen of the Apocalypse*

Food stores were vulnerable, and with each growing season came the risk of crop failure from wind, hail, drought, or infestation. All too frequently, crop failure brought hunger, for surpluses were rare; primitive preservation often was no match for the rot and rodents that consumed what might have accumulated from past harvests. Even quite localized failures could bring famine, for the torturously slow movement of oxcarts and wagons on rutted rural roads provided neither the speed nor the capacity to bring relief, if relief were to be had.

City inhabitants were no better off than the agrarian population. Agricultural setbacks in the surrounding heartland caused prohibitive prices that only the aristocratic class could hope to pay, and malnutrition and worse were the inevitable result. In what are now prosperous countries, entire communities were nearly wiped out by famines well into the eighteenth century. The potato blight of the 1840s brought famine, devastating Ireland, and inadequate food supplies continue to visit the world's poorest countries.

Malnutrition made worse the pestilence that haunted the lives of our ancestors even in the best of times. The ravages of childhood—dysentery, scarlet fever, tuberculosis, and measles—caused many to die before adulthood. These and other adulthood maladies took a continuing toll. Most feared were plague, cholera, typhoid, and other epidemics that periodically swept the countryside, cutting down young and old indiscriminately. Pestilence thrived in the poverty and squalor that once characterized the existence of the mass of humanity. Open cesspools and water supplies contaminated by human waste, rats, fleas, lice, and mosquitoes made fertile grounds for the spread of contagious disease.

And if conditions in the cottages of agricultural laborers were bad, urban life was worse. In the crowded and unsanitary cities, refuse was dropped from upper-story windows and the streets served as open sewers. In London alone, 600,000 died in the Bubonic Plague of 1664; epidemics continued well into the nineteenth century, when four cholera outbreaks took 37,000 lives. But London wasn't unique. Only a few centuries ago, all the world's cities required a steady influx of immigrants from the countryside if they were to grow. In their crowded and unhealthy conditions, disease struck down the inhabitants faster than they could give birth.

War brought death too, but most of the fatalities didn't come directly from cannonball or musket fire. Rather, more were caused by the slide into more primitive living conditions. Armies living in unsanitary field encampments were exposed to extremes of weather and were frequent victims of outbreaks of epidemics. The toll of noncombatants was often larger. When the chaos of war disrupted planting or harvesting, destroyed stores of grain, or created bands of refugees, famine and pestilence were sure to follow. Siege warfare, so prominent in pre-industrial times, deliberately starved those crowded within city walls and weakened the malnourished inhabitants to the onset of smallpox, typhus, and other plagues.

Death from famine, pestilence, and war doesn't complete the catalog of risks that our ancestors faced. Childbirth fatalities and occupational hazards—runaway horses, burns, falls, and frequently infected injuries—added to the precariousness of life. Fire swept crowded cities—built from wood and thatch—turning them into flaming infernos and leaving thousands homeless or dead. From the conflagration that destroyed Nebuchadnezzar's Babylon in 538 BC to the Great Chicago Fire of 1871, many of the world's great cities succumbed, sometimes leaving thousands dead. Likewise, the death tolls from earthquakes, floods, and other natural disasters were multiplied by the famine and pestilence that frequently came in their aftermath.

As frightening as present-day risks may seem, most of us live out our lives without suffering injury or death from the technological mishaps that we dread. With rare exceptions, our only exposure is through newspapers, television, or the Internet. In peacetime, the present-day risk of living in an industrialized country in no way compares to the daily dangers faced by those living in the ages we are tempted to romanticize. Our ancestors' life expectancies were 30–40 years instead of the 70–80 years of today. Little more than a hundred years ago, they faced frequent dangers that struck suddenly and unexpectedly, ending life at a rate that is difficult for us to comprehend.

Simpler though they may have been, the technologies of earlier times were more deadly than those of today. Wooden-hulled sailing ships were lost at sea at a frightening rate, and if a horse bolted or a wagon rolled, death could result not only from the spill's impact, but from infections stemming from sepsis, tetanus, or mishandled attempts at amputation. Likewise, the machinery—powered by picturesque waterwheels and windmills—was simple to understand by today's standards, but was dangerous to operate, causing accidental death at a rate that would be unacceptable by modern occupational safety standards. But technology was viewed differently then, with more hope and less dread. If technological advance could bring deliverance from the poverty and the accompanying scourges of times past, its potential benefits justified risks that today would seem intolerable.

* * * * *

From the beginnings of industrialization onward, engineering has played an essential role in the conquest of disease. Cities remained as crowded and unsanitary as in previous centuries, and the influx of poorly paid factory laborers made conditions worse. The medical profession had no cures for the diseases that plagued the population, but as the nineteenth century progressed, medical investigators began to connect the unsanitary conditions with outbreaks of cholera and other diseases. They gained a better understanding of how disease was transmitted. Even before the germ theory was completely developed, it became evident that poor sanitation, particularly the widespread contamination of drinking-water supplies by sewage, was a source of many health problems, most prominently the cholera epidemics and outbreaks of typhoid fever. Major cities formed public health authorities, and political support grew, allowing the engineering of public-health infrastructures to take root.

In the mid-nineteenth century, engineers developed systems for removing solid contaminants from water. They filtered it through sand and made use of organisms that fed on the impurities. The construction of reservoirs and piping led to systems capable of delivering water free of contaminants to city dwellings. An even greater challenge was building systems for the collection and sanitary disposal of sewage.

Public-works projects in London, Paris, New York, and elsewhere gave birth to elaborate systems of pumping engines, sewer pipes, and settlement tanks to eliminate the waste. In London alone, engineers constructed more than a thousand miles of sewer lines, pumped the waste away from the city, and transformed the Thames from a stinking sewer in the mid-nineteenth century to a river where fish could live just a decade later.

By engineering supplies of pure water, the outbreaks of cholera that plagued the mid-nineteenth century disappeared, and typhoid fever and other diseases caused by waterborne organisms were brought under control as well. The death tolls from other diseases dropped with the improved nutrition and sanitation that rising living standards brought. Ridding homes of flea-carrying rats eliminated outbreaks of bubonic plague, and with the elimination of lice, typhus disappeared. Later, the draining of mosquito breeding grounds and other measures caused malaria and yellow fever to retreat from temperate climates. Tuberculosis was fought by organized efforts to clear the thousands of tons of bacteria-harboring horse manure that accumulated daily on the streets of major cities. Add to these victories the blossoming medical advances that followed in the twentieth century—effective immunization against infectious disease and antibiotics to counter bacterial infection—and the fear of premature death from life-threatening infectious disease in industrialized countries receded from the public's consciousness.

Early in the nineteenth century, the introduction of steam engines compact enough to power riverboats and railroads made a major contribution in vanquishing famine—foodstuffs could now be transported to the afflicted with unprecedented speed. Before the railroads, local crop failures often led to famine: traveling over primitive roads, wagons pulled by horses or oxen weren't able to bring relief with sufficient speed and quantity to relieve suffering. Nor were the horse-drawn barges that plied eighteenth-century canals up to the task of relieving food shortages in isolated villages and towns. But with the railroads came new dangers.

Most early rail lines consisted of only a single track, and the primitive communications systems made collisions inevitable. Before telegraphic communications and signaling lights were developed, miscommunicated or delayed schedules, darkness, fog, and other visual hindrances that interfered with the ability to see a stalled or oncoming train resulted in collisions, causing large numbers of deaths. Poor brakes and crude coupling devices compounded the severity of accidents, and wooden coaches heated by stoves and illuminated with gas lamps frequently turned into infernos.

As trains became faster, heavier, and more numerous during the second half of the nineteenth century, accidents grew more frequent and often more deadly. Moreover, collisions were not the only problem. Heavier locomotives brought increased wear to the rails, roadbeds and bridges; in the late 1860s, bridges in America alone were collapsing at a rate of more than 25 per year. It was these cataclysmic accidents—some taking more than 100 lives—that brought political pressure for stronger action by the railroads and for heightened government regulation. Railroads could no longer be run as if they were stagecoach lines.

Government intervention and engineering advances reduced the risks associated with rail travel. Improved telegraphy and electrical signaling systems, along with two-line systems, reduced head-on collisions. The Westinghouse air brake replaced ineffective hand brakes; improved couplings and standard gauges reduced derailments. Metal supplanted wood, and electricity replaced gas lighting in passenger coaches, greatly decreasing the number of fires. Steel rails, improved structural engineering, and building codes significantly reduced derailments and railroad bridge collapses.

The risks also must be seen compared with the even larger occupational hazards encountered by those who constructed the infrastructures of industrialized society. For example, fewer passengers died in train wrecks than did railroad workers, who perished in frequent construction accidents as they laid track, dug tunnels, and built bridges under the worst of working conditions. Accidents in mining, manufacturing, and other industrial sectors exacted a sizable human toll as well. But frequent as they were, such accidents most often took only one or two lives at a time. Then, as today, accidents that had only one or two victims drew much less publicity and elicited much less political pressure to reduce their frequency. Hence, these accidents, rather than disasters, tended to account for the majority of the death tolls.

Nor was it only accidents that contributed to the technological risks of industrial development. More subtle threats came from adverse health effects attributable to pollutants and contaminants at home and in the workplace. The coal smoke breathed by urban dwellers often approached intolerable levels, contributing to asthma, bronchitis, lung cancer, and other ailments. The air breathed by miners was worsened with the introduction of high-speed cutting machinery. Mercury, lead, and other toxic materials were ingredients in many manufacturing processes, but few if any precautions were taken when working with them—their dangers were often not even recognized.

Later, increased productivity due to farm mechanization played an even greater role in lifting living standards and vanquishing hunger's threat from the industrialized world. But farm mechanization also introduced new risks: the occupational hazards of being crushed or having limbs severed by the machinery, and the danger of explosions in dust-filled grain elevators. But in proportion to the quantities of food produced, these risks were less than those of more primitive farming methods, and they too were reduced as the engineering of agricultural machinery matured. The dangers of steamboats and railroads, of farm mechanization and other technologies, didn't result in the abandonment of their development, though risks were substantial in the early stages. The new risks were less than those encountered in attempting to navigate the same rivers by sail or raft, in attempting to make the same overland trips by stagecoach or covered wagon, or in carrying out many other necessary tasks using what was available before industrialization. The benefits of lifting the standard of living were far greater than the considerable human costs of technology's advance.

In rich counties or poor, war is the one enduring menace that technology has done little to dispel. Military technology advanced—that is, it became more deadly—through the nineteenth century and then became even more deadly. From the sinking of the Lusitania to the collapse of the World Trade Center, organized human cruelty—amplified by advancing weapons technology—has been a perpetual cause of suffering and death into the twenty-first century. But to the extent that peace has been maintained, life in industrialized democracies has become increasingly free of the ancient scourges visualized in Dürer's woodcut of *The Four Horsemen of the Apocalypse*. However, technology continues to present new challenges, and assuring that its benefits outweigh the costs of its risks is a never-ending task.

* * * * *

Over the last century, the accelerating pace of technological advance has brought many benefits to society. But each new technology is to a greater or lesser extent an adventure into the unknown. Accompanying some technologies are new risks that are recognized, understood and brought under control only after frightful events cast doubt on whether the benefits are worth the risk. Two very different examples from the mid-twentieth century illustrate the challenges faced by technologists as they attempt to bring benefits to society without creating risks that are unrecognized and thus uncontrollable.

The jet engine was developed during World War II; after the war, the British aerospace industry sought to use jet propulsion to power a revolutionary new airliner—the de Havilland Comet. With its four jet engines, the Comet flew twice as fast as, and at twice the altitude of, a propeller-driven airliner. Not only did the Comet cut flying times in half, it also cruised above rough weather, thereby increasing passenger comfort. The aircraft also was sleek, with its engines embedded in the wings and fashionable rectangular windows replacing the traditional round ones.

The aircraft's path-breaking performance captured worldwide praise following its introduction. Flights sold out far in advance, and international airlines lined up to order the new aircraft. Then tragedy struck: following takeoff from Calcutta, a Comet blew apart in midair as it climbed toward its cruising altitude, instantly killing all aboard. An investigation followed, but uncovered no apparent cause. Authorities concluded that the freak nature of a monsoon thunderstorm was the cause, or possibly terrorism, but they didn't take seriously the possibility of a flaw in the aircraft's design. A year later, a second Comet broke apart over the Mediterranean after taking off from Rome. Again, terrorism was thought to be the most likely explanation. But, when a few months later a third Comet and its occupants suffered the same fate, the aircraft was grounded, and a thorough investigation commenced to try to explain the catastrophic structural failures that plagued the aircraft.

Investigators were determined to pinpoint the precise cause of the failures—without that knowledge, redesigning the aircraft to rule out further catastrophes would be problematical, if not impossible. Authorities retrieved two-thirds of the wreckage from the second disaster and reassembled it for detailed study. They performed theoretical analyses, tested many scale models in wind tunnels and flew multiple experimental flights. The test that finally revealed the cause of the disasters consisted of submerging the fuselage of a Comet that had completed more than a thousand flights in a huge water tank. The tank was filled and flushed though many cycles, simulating the pressurization-depressurization cycle that the aircraft's cabin undergoes with each flight. After 2,000 cycles, a tiny crack suddenly formed at a corner of one of the windows and spread with lightning speed, destroying the fuselage.

The cause was metal fatigue, the same phenomenon that causes a paper clip to break if you bend it back and forth often enough. Metal fatigue had certainly been considered in earlier design calculations, but knowledge of it had not progressed to a point that would prompt the Comet's designers to properly account for the stress concentrations at the sharp corners of the rectangular windows—concentrations that caused microscopic cracks to form and grow unnoticed, and then erupt, bringing down the aircraft long before it had completed enough flights to justify its retirement from service.

With that knowledge, the aircraft could have been redesigned using oval instead of rectangular windows, thereby eliminating the fatal stress concentrations. But a revival wasn't attempted because the aircraft's notorious history made it unlikely that airlines would buy, or passengers fly, the redesigned Comet. It was left to its competitor, the Boeing Company, to successfully launch the age of commercial jet aircraft a few years later.

Other unidentified risks of new or rapidly advancing technologies may be more insidious. They may not express themselves immediately in the form of identifiable accidents, forcing investigators into action to eliminate the danger. Rather, their adverse consequences may appear as severe health problems, but only years after the technology's introduction. And if the risk is not recognized, many thousands may be exposed and harmed. The early use of X-rays serves to illustrate this danger.

In the 1950s, the adverse effects of radiation overdoses were yet to be appreciated. Thus, in doctors' offices and dentists' chairs, where X-ray machines were valuable new diagnostic tools, little concern was given to the exposure that patients received. Even worse, unattended machines in shoe stores allowed customers to X-ray their feet to examine shoe fit, and children would sometimes play on these machines, exposing themselves for extended periods of time. But the most dire consequence of exposure to radiation turned out to be from the ill-advised uses of radiation therapy to treat children and adolescents for inflamed tonsils, adenoids, acne, and more. These therapies employed intense beams of X-rays, some of which inevitably fell on the neck and exposed the thyroid gland, which was later understood to be extremely sensitive to damage from ionizing radiation.

The exposure of thousands of children's and adolescents' thyroids to excessive amounts of this radiation caused malignancies to develop; twenty years after they were exposed, an epidemic of thyroid cancer ensued. Fortunately, thyroid cancer rarely results in death if diagnosed early and properly treated. In the intervening years, standards were strengthened, and radiation has become a mainstay of medical technology. X-rays and CT scans revolutionized the ability to diagnose injury and disease, and radiation therapy is an essential component of many cancer therapies. However, its history provides a cautionary tale about the necessity of carefully examining potential long-term effects of emerging technologies before they become widely used.

Many additional safety precautions have become law since the mid-twentieth century. At that time, there were no seat belts in our cars, no smoke detectors in our homes, and no sprinkler systems in the hotels where we stayed. Efforts are now more rigorous to identify and deal with risks that may emerge as technology advances at an ever-increasing pace. Compare the examples of two jet airliners: the Comet's failings in the 1950s and the problem encountered with the Boeing 787 Dreamliner today.

As discussed above, three Comet flights blew apart in mid-air, killing all aboard, before safety authorities grounded the aircraft to identify and eliminate the source of the disasters. In contrast, use of lithium-ion batteries in a 787 caused a fire to break out in the battery compartment while the aircraft was parked on the ground. When smoke coming from the battery compartment forced the emergency landing of a second 787 while it was flying over Japan, air safety authorities immediately grounded the fifty Dreamliners then flying to study the causes of the fires and to redesign the battery systems to eliminate the possibility of future safety hazards. In contrast to the Comet half a century earlier, safety authorities mandated that the 787s be grounded, even though the battery problems had caused no fatalities, and, arguably, even the danger of a crash was not imminent.

Likewise, the failure to control X-rays and other radiation a half-century ago allowed their unrestrained use in shoe stores, factories, and medical procedures—some of which had not even been proven to be effective—led to unacceptable exposure to workers in a number of occupations, to medical personnel and their patients, and in some situations to the public at large. All uses of radiation and radioactivity are now highly regulated. Unjustified uses, such as for fitting shoes in stores, are banned; the exposure allowed for X-rays and other commonly used medical procedures has been greatly reduced. And radiation therapy is limited to procedures in which there is evidence that the benefits substantially outweigh the risks.

* * * * *

Despite the tightening of regulations over the last half-century and the increased emphasis on safety, the public's fear of technological disasters shows signs of growing. This fear is kept at the forefront of the public's attention by daily media saturation, much of it emphasizing accidents and other technological mishaps, often with the most bizarre and least likely stories garnering the greatest attention. Books also feed the dread of technology with titles such as *Inviting Disaster*, *Normal Accidents*, and *Why Things Bite Back*, which catalog past disasters and emphasize that even worse ones are likely to come.

But is technology, in fact, becoming more dangerous? Statistics would show that the safety of technology is improving. Per-mile-traveled death rates are in decline, whether that travel is by car, train, or commercial airplane. Workplace safety is improving, fires take fewer lives than in past decades, and with few exceptions, other accidents caused by technology's shortcomings show declining numbers of deaths and injuries. Delayed health effects are more difficult to gage than accidental deaths, which can be counted immediately; even when this is taken into account, evidence doesn't indicate that the situation is becoming worse.

So if technology is not an increasing threat to public safety, why the unrelenting concern with its dangers? Surely its growing complexity and the public's difficulty in understanding play a role. But there is another explanation. It relates to the rapid reduction in deaths caused by disease.

Over the course of the twentieth century, improved sanitation, inoculation, antibiotics and the increasing effectiveness of medicine in general have conquered many of what once were the leading causes of death. In industrialized economies, deaths from communicable diseases, infection, and complications in childbirth have plummeted. Consequently, life expectancy has increased greatly.

In the United States, for example, life expectancy has increased by more than 30 years over the last century. The medical challenges and leading causes of death are now primarily those of aging, most prominently heart disease, stroke and cancer. Behind these, accidents—most of them related to technology in some way—have become the fourth leading cause of death, and the leading cause for those between 1 and 40 years of age. This does not imply that the death rate due to technological mishaps has increased over the past decades. On the contrary, that death rate has dropped substantially, but in industrialized countries, death rates due to malnutrition, disease, infection and other causes have dropped much faster.

Not surprisingly, with other causes of death decreasing more rapidly, accidents and environmental contamination have become a growing concern of public officials and citizens alike. Since the most widely reported mishaps frequently involve shortcomings, malfunctions, or misuse of the technologies that permeate industrial societies, their risks loom larger in the public's mind. As a result, the historic role that technology has played in overcoming obstacles to humankind's well-being has tended to fade from the public's collective memory and many of the benefits that it provides are taken for granted.

Do the gains that have been made in increasing technology's safety mean that less attention need be paid to its risks? Absolutely not! Engineers, corporations, and public officials should not become complacent. Reducing the thousands of fatalities resulting from the technologies that pervade modern economies is an ongoing struggle, and identifying and reducing the risks that might accompany newly emerging technologies is essential if technological risks shouldn't increase.

Indeed, public pressure to reduce the frequency and severity of accidents and other adverse effects that have stemmed from technology's use has increased markedly over the past decades. Responding to that pressure, those who design, build, operate, and maintain products, processes, and services upon which our economy depends have sought to create technology that provides a greater degree of safety, even as they have wrestled with the often-conflicting goals of decreasing the costs and improving the performance of their creations. But inevitably, as the resulting technology comes into use, the question is: Is it safe?

This seemingly simple question applies whether buying a lawn mower, toaster, or the latest-model automobile. It is asked when deciding to travel by car, rail, or air, and when deciding whether to stay in a high-rise hotel or cross an aging bridge. "Is it safe?" is foremost in the public's mind when plans are made to build a new power plant, refinery, or offshore drilling rig. But ask a design engineer, plant manager, or government regulator that question, and a complex array of interlocking issues spring immediately to mind. To them, a textbook definition such as "safety is the state for which risks are judged to be acceptable" only scratches the surface: this definition must be interpreted and applied to myriad decisions that design engineers make in creating and building new technology. It must be built into the rules and procedures by which industrial establishments are run. It influences how regulators interpret and enforce the safety rules and regulations by which political processes have determined what level of risks are acceptable to the public.

If the question is rephrased as "Is it absolutely safe?" then the answer must be no. No technology can be made totally risk-free. No aircraft can be guaranteed not to crash, no ship not to sink, no building not to collapse, no pipeline not to explode, and no electrical appliance not to cause harm. Thus, the question for citizens and engineers alike is not simply "Is it safe?", but rather the twofold question: "How safe is it?" and "Is it safe enough?"

Although it is difficult, we may strive to answer the first of these questions through objective analyses of the creation and use of technological systems. For mature technologies—those that are now evolving more slowly with time—we can gather statistics on accident and death rates incurred from their use. Thus, authorities can compile fairly accurate measurements showing how safe travel by automobile or airliner is, and indicating the risks encountered using industrial machinery at work or common appliances at home. But for new or rapidly evolving technology, the problem of measurement becomes much more difficult, if not impossible, since its use hasn't been extensive enough to gather statistics on its risks. Engineers must concentrate instead on identifying whatever potential hazards may lurk in their designs. They must eliminate or reduce them to manageable proportions before their technology becomes available for public use. Only as experience accumulates with the technology's widespread use can statisticians make reasonable estimates of how safe it actually is.

"Is it safe enough?" This is a more difficult question to answer, even for more mature technologies, whose idiosyncrasies are understood and statistics have been gathered regarding the fatalities or illness stemming from their use. And it is more difficult yet for newer, more innovative technologies, for which there is still great uncertainty in answering the question of how safe it is.

The question "What is safe enough?" constitutes a value judgment that requires weighing risks against benefits and depends both on our individual dispositions toward safety and those of the society in which we live. Answering it often brings about internal conflict between our innate fears—irrational as they may sometimes seem—and our attempts to logically examine the nature of the risk and put it in perspective by comparing it to the other risks we face. This is a dichotomy that plays out in society as science, psychology, politics and economics interact in attempts to decide what is safe enough.

* * * * *

Attempts to deal with the question of how safe is safe enough have evolved greatly over the past century as technology has become more complex and pervasive, and the public's demands for safety more pronounced. A century ago, professional codes of good practice spread by engineering societies and other professional organizations served as a basis for building codes, consumer product certification, and more; they continue to be incorporated into a great many of the ordinances and laws dealing with safety today.

With the increasing complexity of technology, public pressure has grown to create more formal procedures for assessing its health, safety and environmental impacts and determining whether the public is adequately protected from its risks. The United States provides a prominent example of how public pressure has resulted in increased levels of regulation. A multitude of federal regulatory agencies has come into being. The first, which would eventually deal with technological risks, was created as a result of the 1906 Pure Food and Drug Act and assumed its present name, the US Food and Drug Administration (FDA), in 1930.

Public pressure to improve safety and limit damage to the environment resulted in the creation of many new agencies in the 1960s and 70s: the National Highway Transportation Safety Administration, the Environmental Protection Agency, the Consumer Product Safety Commission, the Federal Aviation Administration, the Nuclear Regulatory Commission and many more.

Through the regulations put into effect by these agencies, and those of state and local governments as well, society determines how safe is safe enough. Elected officials reflect public opinion in passing and attempting to enforce legislation to deal with technological risks. Their actions are often prompted by public outcry following a disaster—an airliner crash or refinery explosion—that has resulted in spectacular destruction and substantial loss of life.

Legislation also arises from legislators' constituents lobbying in response to extensive media coverage of an ongoing pollution problem suspected to be a threat to public health, or to the publicity given to particularly poignant accidents, such as those involving children. No matter the source of political pressure, neither legislators nor their staffs are likely to have the expertise to spell out the regulations in detail. Thus, what is passed into law typically is couched in quite general terms, leaving an appropriate government agency to work out the specifics. Corporations, design engineers, operating personnel, enforcement officials and the public must adhere to these specifics.

The regulatory agencies at the national level arguably are among the more sophisticated practitioners of translating legislated mandates into detailed rules to govern safety practices. They employ staffs thoroughly versed in the methods of quantitative risk assessment. These approaches, such as cost-benefit and cost-effectiveness analysis, attempt to cast risks in objective terms, such as numbers of lives or life-years lost, numbers of emergency room visits, costs of medical treatment and the like. They strive to combine scientific understanding with statistical analysis to determine what is safe enough. But the public may not be receptive to these methods, for they invariably include difficult-to-understand probabilistic concepts, and the required objectivity often seems callous, such as putting a dollar value on human life in order to weigh costs against benefits.

The risk analysts' methods frequently come head to head with the public's vastly different perceptions of risk and with industry's financial concerns in the public hearings that are a required component of the rule-making procedures of most agencies. In such processes, citizens' voices are heard as they demand that government respond to their fears and unease, even if "risk experts" feel that those fears are irrational or unjustified. Lobbyists from the affected parties may also intervene. They may argue that the proposed rules are unworkable, that they may cause unintended or undesirable consequences, or that they may require excessive costs to implement.

Regulatory processes are most often messy and contentious, but they are how quantitative assessment, public perceptions and economic interests come together in determining how safe is safe enough. In general, the results are positive. The hearings may undergo long delays when those opposed to more stringent regulation demand more study before action is taken or use other stalling tactics in the hope that, as time passes, public demand for action will lose its intensity. Negotiations invariably result in compromises between the public, who originally put pressure on their elected officials for change, and the corporations, government bodies, landowners, and others who have financial, aesthetic and other vested interests that may be at odds with the drive for stricter safety laws.

In the end, the agency heads or commissioners must settle on a solution. They are usually political appointees, subject to biases and pressures stemming from their political leanings and professional backgrounds. They must approve the rules written by the agency staff to represent the compromises reached; these are often at odds with the quantitative risk assessments performed within their own agencies—and in the case of commissions, approval is often made by split decisions. But even then, the agencies may be overruled. Affected parties may take the agency to court, and in some cases the courts have overruled regulations. Subjected to constituent or lobbying pressure, Congress may also overrule the agency with amended legislation, or the

President's administrator of the White House Office of Information and Regulatory Affairs may modify or delay implementation of the regulations that emerge from the foregoing process. It's a fascinating story, the saga of how a society's political, economic, engineering and social interests interact to reach the decisions that determine the level of risk to which we are exposed.

CHAPTER 2

Why Things Fail—The Bathtub Curve

On the afternoon of January 15, 2009, US Airways Flight 1549 took off from New York's LaGuardia Airport and headed for Charlotte, North Carolina. Three minutes into the flight, as the Airbus A320 made its initial climb, the first officer spotted a flock of Canadian geese ahead on a collision course with the aircraft. An instant later, the windshield went brown as pilots heard the thuds of colliding geese. Worse, birds sucked into both of the aircraft's engines caused the aircraft to lose the thrust needed to keep it aloft. The plane lost altitude too quickly to make it back to LaGuardia or to Teterboro Airport in New Jersey, making the likelihood of disaster imminent. If the plane came down among Manhattan's high-rise buildings, not only would the 155 aboard lose their lives, but so would many more in the densely populated city.

Thankfully, as the windshield cleared, the captain turned the aircraft to the south, skillfully glided it 900 feet over the George Washington

Figure 2.1: The Hudson River landing of US Airways Flight 1549, *Associated Press*

Bridge and within three minutes of the bird strikes, made a smooth landing on the Hudson River opposite midtown Manhattan. The captain and crew ushered the passengers out the emergency doors, some onto the wings, and others onto the floating escape chutes. Four minutes later, the first rescue boats arrived to ferry the passengers ashore. The extraordinary actions of the captain and crew averted disaster and saved the lives of the 155 passengers and crew. The landing became known as the "Miracle on the Hudson."

The cause of the accident may be viewed as the failure of the aircraft's two jet engines. The failures did not depend on the age of the engines; the result would have been the same had the engines been brand new or if they had been nearing the end of their useful lives. In this sense, they may be classified as random failures; they could have occurred at any time. The engines failed because they encountered an environment that they were not designed to withstand without shutting down: the entry of large birds.

Jet engines are designed to survive severe environments, but how severe should those environments be? There have been numerous cases in which birds have been sucked into engine intakes; in fact, engineers follow FAA regulations in designing engines to survive bird strikes.

These regulations require the designs to counter the effects of the entry of light, medium-weight, and heavy birds. Strikes from small and medium-sized birds are most frequent, and indeed more than one bird may strike an engine if the aircraft encounters a flock. Thus, the regulations require that an engine retain most of its thrust following such strikes. In some tests, the engines must continue to function following simultaneous strikes by as many as seven small birds. For large birds—the size of a Canadian goose—the regulation states that the engine must safely shut down without catching fire or otherwise endangering the aircraft. The regulation thus assures that an airliner can land using another engine should such a large bird strike occur.

But what about a less likely event? What if Canadian geese hit both engines, as they did with flight 1549? Both engines shut down safely as the regulations require, but without propulsion, the aircraft lost its ability to fly. Should the regulations be strengthened to stipulate that the engine must continue to operate after sucking in a Canadian goose? Even then, there is still a risk. Geese travel in flocks, and more than one may be ingested into an engine. Should the requirement then be that the engine shut down safely, or that it continues to operate?

At some point, the difficulty in designing jet engines would be beyond the capabilities of existing technology; if such engines could be produced, their weight and expense would preclude their use in commercial aviation. The probability of multiple large bird strikes occurring must be taken into account, and it must be weighed against increasingly vigorous efforts to decrease the probability of bird collisions by using a variety of methods to deter birds from nesting or flying near busy airports. Authorities face difficult decisions about at what point the probability of a cataclysmic bird strike is small enough that engine designers should not be required to provide protection against it.

Jet engine designers must anticipate other adverse conditions as well. Volcanic ash is one of the more serious. Nearly 100 aircraft have been damaged by ash pulled into their engines, and some have narrowly averted disaster. In 1989, a Boeing 747 was caught in a volcanic cloud from an eruption of Mount Redoubt in Alaska. Ash shut down all four engines, and the 231 passengers experienced nearly five minutes of terror as the KLM airliner lost more than two miles of altitude. Because the disabled engines cooled as the plane descended, the pilots were able to restart them and land the plane safely .

Fortunately, when the Icelandic volcanic eruption took place in 2010, disrupting air travel over Europe and the North Atlantic, warnings were sufficient that (although some aircraft suffered engine damage) no engines shut down. Nothing could be more frightening than what was experienced by the occupants of KLM's flight over Alaska.

Improved methods for allowing pilots to avoid volcanic plumes will lessen such dangers, but should jet engines be made more robust, allowing them to encounter thicker volcanic clouds without failing? If so, how much thicker? The questions raised are even more complex than the bird strike problem, where shooting dead geese into running engines allows engineers to judge the success of their efforts. We know that volcanic ash is a glassy particulate that melts at the high temperatures inside jet engines and prevents the cooling of the turbine blades.

But to improve engine design, a much deeper knowledge of the ash's characteristics and how it interacts with the engines must be gained. This would require more than laboratory studies. Aircraft—without passengers—would need to be flown into volcanic clouds to determine the characteristics and density of the ash and relate it to the nature and extent of the engine damage. But as design improvements are proposed, verifying that they will work is challenging. Our understanding is not near the point where computer simulations can be relied upon to predict how the improved engine will behave in a volcanic cloud.

Prototype engines can be built, but how will they be tested? Do the engineers have to wait until the next volcanic eruption takes place and fly a plane equipped with the new engines through it to see what happens? There would be unacceptable delays, waiting for the next eruption to take place. Likewise, design is a test-and-fix, test-and-fix process when new and poorly understood conditions are in question. Thus, the properties and behavior of volcanic dust would need to be simulated with sand or other material so that engines could be tested on the ground and go through cycles of analysis and improvement before they are put to the test by flying them through a volcanic plume. But even after the tests are completed, the question remains: Is the engine safe enough, or should the expensive quest continue to allow the engine to survive under even more severe volcanic conditions?

* * * * *

Unlike machinery produced in large numbers, such as jet engines, the impossibility of building prototypes to test one-of-a-kind structures severely limits engineers' capability to predict a structure's ability to avoid random failures caused by severe, if improbable, conditions. The consequences of natural disasters head the list of challenges, whether they are floods, tornadoes, avalanches, earthquakes, tsunamis, or volcanic eruptions. How severe a natural disaster should a dike, dam, building or bridge be built to withstand? Should it be able to survive a natural disaster expected to take place once in 100, 500, or 1,500 years? These are not questions whose answers fall strictly within a framework of scientific analysis. A great deal of judgment must come into play because the answers cover psychological, economic, philosophical, and even theological considerations. They must reflect the public's perceptions of risk and its aversion to it.

Even after a number has been specified for the frequency of occurrence beyond which a structure may be expected to fail or be severely damaged, a number of challenges remain. First, what will the magnitude of the load accompanying that event be? Second, how can we assure that the structure will withstand that load?

Consider earthquakes: If we say that a building should survive a once-in-a-thousand-year earthquake, how severe is that quake? Statistical models have been developed relating the severity of quakes to their frequency of occurrence—the less frequent the quake, the greater its magnitude. But the calculation of such probabilities is imprecise, and the less frequent the events, the more the estimates of the probabilities come into question. At a given location, historical data likely has not been collected for long enough to make the estimates very precise. Moreover, earthquake damage is not related directly to the magnitude of the quake—usually measured on the Richter scale—but to the severity of the local ground shaking. That shaking depends not only on the quake magnitude, but also on the depth of its epicenter, its distance from the structure, and the geological structure of the earth's crust between the quake's epicenter and the structure.

Once these considerations are compiled into an estimate of the severity and nature of the shaking that a structure must be designed to withstand, much remains to be done. Sites where recent earthquakes took place must be examined, and the nature and extent of the damage to similar structures analyzed. Laboratory tests are set up with massive vibration tables constructed with the capability to reproduce the shaking recorded on seismographs during severe earthquakes. Then sample beams, girders, connections and other parts of structures are subjected to shaking, and their responses and damage analyzed.

Investigators set up computer simulations and test their accuracy compared to experiments and the damage incurred in actual earthquakes before engineers use simulations to predict the earthquake resistance of newly designed structures previous to construction. Earthquakes are but one example of the severe randomly occurring environments that society and its engineers must deal with: how robust its creations must be, how severe, albeit improbable, the events they must withstand.

Random failures, whether caused by bird strikes or earthquakes, take place when technology is exposed to environments that it cannot withstand, irrespective of its age. These failures can happen for a number of reasons. There are situations, though rare, for which it is next to impossible to design against failure. Aircraft are not designed to survive midair collisions, nor are automobiles built to withstand total submersion. Other failures result from misuse of products, whether inadvertently or through ignorance. A chainsaw will destruct if used to cut metal, as will a garbage disposal subjected to metal refuse. In some cases, warnings placed on products can alert buyers to inappropriate use, but to most people, such warnings appear laughable, since they seem obvious to nearly all who are familiar with the technology.

Nevertheless, engineers must design in such a way that inappropriate use of a product is as foolproof as possible. However, an even more important concern is to determine the severity of the environment that the product will be in and to design the product to survive that environment without failing. The common cell phone offers an example: The design should be sufficiently rugged to survive a fall off a table, but what about from an even greater height or down a flight of stairs? The phone should survive being left in the rain, but what about falling into a bathtub and left there for a while? The phone probably should be able to survive being stepped on, but by how heavy a weight? The "what-ifs" are endless. Even where there is little or no safety concern, designers must determine the limits of environmental stress that their creations should be expected to survive, convert those limits to specifications, and ensure that their products can meet them.

* * * * *

Too frequently, structures, machines, and other forms of technology fail early in their expected life, for example in the first months, even though engineers designed them to last for years. Such early failures—often referred to as infant mortality—occur when defective manufacturing processes or construction practices fall short of creating technology as strong as is detailed in the designers' plans and specifications. The resulting defects are likely to cause failures shortly after the technology is put in use, even though the stress that it is subjected to is less than it was designed to withstand.

In mass-produced products, defects may result from multiple issues. Among the most important are the human variability of the production workers and the inability of factory machinery to precisely control material properties, surface finishes, or critical dimensions. Automobiles and computers offer contrasting examples of the imperfections that arise and of efforts to control or eliminate defects that lead to early failures.

In earlier days, automotive production was much more labor-intensive than now. Consequently, human factors played a larger role in defects arising in the assembly process. It was sometimes said that you didn't want to buy a car that had been assembled on Monday because the workers were less attentive after returning to work after the weekend. Training and attention to work rules that assist laborers in remaining focused throughout their work shift are essential in preventing errors arising from carelessness, boredom, or distraction. A good deal of laborers' work has been automated, but this doesn't lessen the need for vigilance in quality control. The control settings on robots may drift with time, cutting tools dull, and material suppliers fail to conform to standards. Continual monitoring and inspection remain essential.

Running an exhaustive set of tests on each automobile before shipping would greatly reduce the probability that any defect that survived the manufacturing process would remain in the car. But such testing would add unacceptable cost increases to the purchase price. Instead, manufacturers run only a limited number of tests on every auto coming off the assembly line. They then perform more comprehensive testing only on a scientifically chosen sample of the autos. Engineers trace the defects found back to their origin in the manufacturing process and correct the process. At the same time, they must isolate other autos that may have come off the line with the same defect, and implement a procedure to correct it before the cars are sold.

From Leonardo da Vinci's fifteenth-century machines to the electronics of the twenty-first century, manufacturing technology has been pushed to the limit in the race to reduce costs and achieve ever-increasing levels of performance. Such is often the case in efforts to produce ever-faster and smaller computer chips. Limitations do not arise as much from the ability to design the integrated circuitry as in the difficulties of fabricating those designs in chips that are free of manufacturing defects. As the spacing between transistors and the thickness of the chip's internal wiring are reduced to the scale of nanometers—billionths of a meter—controlling their dimensions is at the limit of manufacturing technology. Consequently, not all of the chips produced are perfect, and the yield of defect-free chips is often substantially less than 100 percent.

Engineers use a number of techniques to assure that the potential for early failures remaining in the chips that go to market is very small. One of these, "shake and bake," entails shaking the chips, since severe vibration and exposing them to high temperatures are likely to cause substandard chips to fail before they are tested for functionality. Likewise, "burn-in," in which the chips are operated at temperatures exceeding those encountered in normal use, will cause most of those containing defects to fail within a fairly short period of time. Thus, chips that have survived such procedures are much more likely to be free of the manufacturing defects that cause early failures.

* * * * *

Whether we're talking about an automobile, computer, or other mass-produced products, designers' calculations are validated by building prototypes and making sure that they work, are strong enough, and can withstand the stress of use without failing. But large, one-of-a-kind structures—buildings, for example—do not have prototypes, making weaknesses more difficult to detect and eliminate. An adage of structural engineering is "you live in the prototype". Other methods must be employed to ensure that a large structure will behave as the designers intend.

Models and component testing supplement calculations, for example, in predicting how a large bridge or other structure will behave. But proof testing is also likely to be employed. Engineers may place loads of concrete or steel on a newly constructed bridge that far exceed the weight of the traffic that it will carry. They make measurements of stresses and deflections to ensure that the bridge will function as intended. And if the results indicate weaknesses, authorities will require reinforcements before public use begins.

Proof-testing methods must, of course, be attuned to the form and function of the technology. A case in point is large ships that undergo shakedown cruises to isolate and correct defects before they are allowed to take on passengers or go into battle. Those one-of-a-kind technologies that can't be effectively proof tested before deployment are more vulnerable to destruction as a result of early failures. For example, NASA's Genesis space capsule, which was designed to collect data on the solar wind, came crashing to Earth because four switches had been installed upside down.

Whether in bridges, ships, or buildings, undetected departures from the architects' designs during construction arguably are the most common causes for early failures of large one-of-a-kind projects. Such changes may be made to reduce costs, facilitate scheduling, or compensate for unanticipated soil or related environmental conditions. If the alterations weaken the structure or otherwise make it susceptible to failure even under anticipated usage, a genetic defect may be said to be imbedded within it, and an early failure—or infant mortality—thus results. Such early failures took place in Kansas City in 1981.

The multistory atrium of the Hyatt Regency Kansas City was spectacular, with skywalks suspended from the ceiling two, three, and four stories above the lobby. A year after the hotel opened, 1,600 people gathered in the lobby and on the skywalks to witness a dance competition. But as the crowd gyrated to the rhythms, tons of steel, glass and concrete from the highest skyway, suspended 45 feet above the lobby, came crashing down onto a lower skyway and then to the crowded lobby below. Of the partygoers, 114 died, and many more were injured.

The accident was caused by deviations from the architects' design, made by the building contractors to circumvent construction difficulties. They modified the method by which the steel suspension rods were connected to the walkways. Although contractors consulted the structural designers, no one went back and calculated the added stresses placed on the structure as a result of the reconfigured connections. In fact, the stresses were doubled, making the connectors barely strong enough to support the weight of the skywalk and certainly insufficient to support the added weight of its occupants in addition to the vibrational stresses caused by the dancing. It was a classic case of an early failure of a structure not built as designed.

* * * * *

No technological design has an infinite lifetime, and aging causes the ability to resist stress to decrease and the likelihood of failure to increase. Exposure to harsh environments accelerates the aging process. Temperature extremes, and especially the cycling between frigid and hot weather—freeze-thaw cycles—aggravate everything from the development of potholes to crack growth in metals, plastics, and other materials. The corrosive effects of salt cause ships and bridge abutments to age faster in saltwater than in fresh. A deadly example of the damaging effects of salt and humidity occurred over Hawaii in 1988.

Following takeoff, an Aloha Airliner reached an altitude of 24,000 feet, where the pressure of the thin atmosphere is much less than of that inside the pressurized cabin. Suddenly, as the aircraft flew above the ocean, a small section of the cabin roof ruptured. The chief flight attendant, standing in the aisle, was blown out the opening to her death; the difference in air pressure immediately ripped away the entire upper half of the cabin's skin, turning the aircraft into a convertible.

Figure 2.2: Aloha Airlines airliner after coming "unzipped," *Associated Press*

Fortunately, unlike the Comet disasters decades earlier, the remainder of the aircraft stayed intact. The explosive decompression, however, caused many injuries among the aircraft's 95 occupants, eight of them serious. The decompression also severed electrical lines, making it impossible for the captain to know whether the nose landing gear was functional. For ten horrific minutes, the passengers clung to their seats, with instructions to don life jackets in case the plane would not make it back to land. The damage was severe, but through skill and decisive actions, the captain and crew brought the plane safely to an emergency landing at Maui's Kahului Airport.

The Boeing 737 was designed so that, even though a rupture took place, it should not have spread; the cabin structure should have retained its integrity. But age had taken its toll. The aircraft had made many short flights, resulting in nearly 90,000 compression-decompression cycles of atmospheric pressure on the fuselage, while the design life of the plane called for only 75,000. But more than that, the humidity and salt of the Hawaii environment had accelerated the effect of fatigue—that is, it exaggerated the repetitive stress cycle that took place with each flight. Corrosion formed around the lines of rivets that held the cabin roof in place. Cracks emanated from the rivet holes and grew at a rate much accelerated by increasing corrosion. Finally, the cracks between rivet holes connected, causing the entire seam to become "unzipped," with disastrous results. This dramatic example of an aging failure had far-reaching effects on aircraft safety, particularly on the frequency and detail required in inspection to uncover signs of corrosion and the initiation of cracking long before they can become a danger.

To some extent, durable materials and thoughtful design can lessen the damaging effects of age, but even in more benign circumstances than encountered by the Aloha airliner, maintenance, repair, and the replacement of shorter-lived parts are central to staving off the increasing rates of failure with age. That's why automobile owners get their cars' oil changed, filters replaced, brakes relined and tires rotated and replaced. Whether aircraft or oil rigs, bridges or buildings, maintenance is essential not only to slow the aging process, but also to know when to declare the end of life and remove a technology from service before harm is done.

Maintenance and repair, however, are only as good as the competence of the organizations and workers who perform them. Faulty maintenance can lead to failure or a shortened life. A friend recalls having tires rotated, but then hearing a rattling sound, stopping and taking off the hubcaps to discover that several of the nuts had sheared off as a result of not being tightened, causing the wheels to wobble as the car moved forward. Fortunately, the problem already became apparent at low speed. Had the car gone onto the expressway, a wheel might have come off, with disastrous results. Unfortunately, far more deadly events have happened due to faulty maintenance. One of the worst took place in 1979 at Chicago's O'Hare International Airport.

Seconds after lifting off the ground, the leftmost of the three engines of a DC-10 detached from the wing, flipping over it, ripping off a three-foot chunk of metal, and rupturing hydraulic and electric lines vital to the control of the aircraft. The plane rolled over on its side and then plunged to the ground within a mile of the runway's end. All 271 occupants and two people on the ground died in the fiery crash.

The National Transportation Safety Board's investigation determined that the cause of the crash was faulty maintenance: jet engines must be periodically removed from airliners to undergo thorough overhaul. But in removing and reattaching the engine of the doomed DC-10, the maintenance crew damaged the pylon, which connects the engine to the wing. Consequently, pylon cracks developed and grew larger with each flight until, on that fateful day in May—six months after the maintenance had been performed— the stress of takeoff caused the weakened pylon to pull loose from the wing and result in the largest loss of life from an aircraft coming down over US soil.

* * * * *

A good deal of commonality exists in the failure mechanisms of all complex organisms, whether they be biological or technological, human or produced by humans. In fact, the earliest studies of failure come from human demography, in which practitioners attempted to determine not only life spans but also the distribution of life expectancy. From these studies came a curve with the shape shown in Figure 2.3: that of a traditional bathtub.

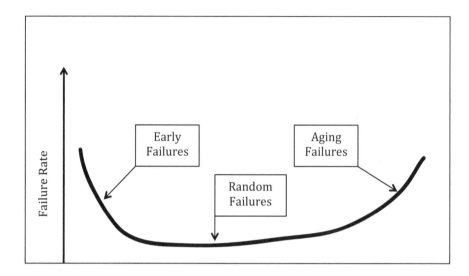

Figure 2.3: The Bathtub Curve

The curve displays failure rate—the death rate in the case of humans—on the vertical axis against time on the horizontal one. The failure rate is a probability that an organism will fail within the next increment of time—a year for example—given that it has survived until that time. Although the bathtub curve first appeared in attempts to understand mortality rates in human populations, its value became apparent in studying technological failures as well, and it is now a ubiquitous tool of reliability engineers.

The bathtub curve decreases sharply at first and then flattens out for an extended period of time before rising again at an increasing rate. Although they interact, different failure modes dominate the decreasing, flat, and increasing parts of the curve. Random failures occur along the entire length of the curve, but they dominate the flat central section. Random failures are characterized by causes that depend little, if at all, on the age of the organism; they result from stresses beyond those for which it was designed to withstand.

In human populations, infectious diseases that were fatal to people of all ages once were the dominant causes of death that had little respect for age. Accidents likewise contributed substantially to the toll. With the conquering of many epidemic diseases in more modern times, random deaths are fewer, but accidents account for a much larger fraction of those remaining. As we saw in the example of jet engine bird strikes, random failures occur in technology when a product or system encounters external stresses that exceed those for which it was designed; aircraft are not designed to survive midair collisions, nor trains to avoid destruction in head-on collisions.

On the left side of Figure 2.3, the failure rate starts out high but decreases with time. In humans, these early fatalities are classified as infant mortality. They result from congenital defects and from the increased vulnerability of children to infectious disease, accidents, and other external attacks until they approach maturity. In dealing with technology, we refer to these as "early failures." As the collapse of the Kansas City hotel illustrates, early failures stem from defects in manufacturing or construction that cause failures early in the expected life span of a product or structure because such defects cause technology to be weaker than it was designed to be.

Aging or wear failures account for the increasing failure rate on the right of the bathtub curve. Again, we find similarities between the frailties that accumulate with age in humans and in technology, frailty that causes exposure to everyday stresses to lead to increased occurrences of failure. Arteries narrow; piping systems corrode. Bones are subject to osteoporosis; technological structures crack, fray, and undergo other forms of deterioration.

Aging and lifespans of biological organisms vary greatly. Elephants live longer than horses, horses longer than dogs, and dogs longer than mice. Likewise, the design lives of technological products vary greatly, and thus also must the time span over which the bathtub curve stretches. Massive structures—buildings, bridges, and dams—may be expected to last for a century or more. Machinery lasts for a shorter time: airliners are kept in service longer than automobiles, automobiles longer than dishwashers, and dishwashers longer than toasters or curling irons.

Where technology advances most rapidly, in computers or cell phones, for example, products typically become obsolete before they wear out, so that the rising failure rate on the right of the bathtub curve is of less concern. But for less rapidly evolving technology, the deterioration that comes with age and associated safety concerns plays a prominent role in design, manufacturing and maintenance decisions.

Since early in the Industrial Revolution, both medicine and engineering have contributed greatly toward reducing the causes of death, thereby pushing human death rates downward and increasing life expectancy. Antibiotics, vitamins, surgery, and more have been beneficial, but so have the technologies for clean water supply, sewage disposal and medical devices, to name a few. The drive for safer, less accident-prone technology also contributes to reducing human death rates and extending longevity.

* * * * *

For the situations discussed thus far, failures have resulted in hazardous consequences. But this is not always the case, and to distinguish between hazardous and harmless failures, we must clarify the difference between safety and reliability. Safety implies the absence of hazards, while reliability is the opposite of failure. A technology is reliable if it is not prone to failure, if it is not likely to stop functioning during the lifetime over which it is designed to last. Some technologies, however, may be reliable but unsafe. A piece of machinery may be virtually failure-free, for example, but if the design does not protect workers from high-speed cutting blades or dangerously hot surfaces, it is unsafe. Conversely, many consumer products are unreliable but quite safe. Such is the case if they are not hazardous to use but break down frequently, but in ways that create no hazards.

If a technology is prone to failures that result in hazards, then it is neither reliable nor safe. For much of the technology discussed in these pages, any failure is hazardous. Any failure of an aircraft's engine, for example, is hazardous. In such situations, safety and reliability are synonymous, and we may use the words interchangeably.

Since no technology will last forever, technologists must give careful thought to how their creations are to be maintained and eventually withdrawn from service. In many situations, a device may be used until it fails. When that is the case, it is imperative that it be designed so that when failure occurs, it causes no harm. Consider as an example the circuit breakers we all have in our homes. Their purpose is to open, thereby shutting off power, if the current passing through it becomes too great. If they fail to open in such circumstances, the faulty wall wiring, appliance, or lamp causing the excess current is likely to overheat and catch fire.

Circuit breakers typically are expected to last decades or more and often are not subject to periodic inspection. Over the years, the breaker's material properties may deteriorate, which may result in a drift in the magnitude of the current required to cause it to open. If the drift is downward, smaller and smaller currents will cause the breaker to open. Harmless failures, caused by inconsequential current levels, will then occur with increasing frequency, causing inconvenience and cost. Such "false-positive" openings, however, result in the need to reset the breaker, but it also worries the owner: Was there really a problem and a need to take action to correct it? Or was the breaker at fault? The aggravation of repeated false positives normally leads to calling an electrician to replace the breaker, and no harm is done.

But what if the drift is upward? Eventually, dangerously high currents could pass though the breaker without it opening; electrical problems would go undetected, and fire might well result. Thus, the breaker would be said to have failed dangerously, allowing "false negatives." Unfortunately, the aging mechanism might not be noticeable to homeowners, even if they occasionally looked into the breaker box, since only electrical testing will reveal the breaker's drift. Therefore, it is imperative for the breaker design to incorporate an aging mechanism that will cause it to fail harmlessly. Otherwise, only an inalterable plan of inspection and replacement will assure the homeowners' safety.

It is best, of course, to design in ways that replace failure mechanisms leading to hidden hazards compared to those that will not. If this is not feasible, some other approach should be taken, such as adding a safety mechanism that will provide an indication that the product is approaching a failed condition, so that it can be retired or repaired before it fails. The home smoke alarm is a familiar example: Battery depletion is the most frequent failure mode, and the solution is a simple control system. Such circuitry causes the alarm to begin chirping well before failure occurs, so owners are alerted that maintenance is required, but they are not frightened by a sudden blast from the alarm, as would occur if a fire had broken out.

In creating any device or system, whether it is a radio or a refrigerator, an automobile or an airplane, the designers must carefully analyze potential modes of failure and determine either how to eliminate them completely or reduce their impact to ensure not only economic success, but safety as well. Methods for reducing early, random, and aging failures differ considerably. Their application also differs from one industry to another, and from product to product and system to system. Failures due to each of the modes may sometimes be inconsequential or result only in inconvenience or economic loss. But when safety is at issue, each of these failure modes can lead to disaster.

* * * * *

Whether in biology or physics, redundancy is invaluable in staving off failure, no matter whether failure is caused by defective construction, random events, aging, or wear. Thus, we have two eyes, lungs and kidneys, and commercial aircraft always have at least two engines and two pilots. In a perfect world, the gains resulting from redundancy would be impressive: if an engine has a failure probability of one in a thousand, the probability of both failing will be one in a million, and of three failing, one in a billion. But in a perfect world, the failures are completely independent of one another, while in the real world they frequently are not. So while redundancies can provide reductions in the probability of failure, they don't come close to the reductions predicted if independent failures are assumed. The redundancy of a two-engine jetliner again provides an instructive example.

Today, all two-engine airliners can fly and land safely on one engine. Moreover, if one fails, the second needs to function only long enough for the plane to fly to the nearest airport. Decades ago, when jet engines where less reliable than now—failure-prone enough that four rather than two engines were required on transoceanic flights—most failures came from problems in the internal workings of the engines; indeed, most were independent failures. But now individual engines are so reliable that the chances of two failing independently are truly minuscule.

The remaining dangers now tend to center on what reliability engineers term "common mode" failures: the simultaneous failures of redundant components from a common cause. Employing redundancy and physically isolating the engines from one another can also reduce common mode threats from malfunctioning sensors, control circuitry and so on. But what if there is bad fuel, or a maintenance technician makes the same mistake on both engines? Here, use of a second principle, diversity, can help. Diversity may mean using fuel from two different suppliers (if the tanks are isolated), or assuring that the same team of technicians doesn't maintain both engines.

Diversity, incidentally, has also been used in the cockpit by not allowing pilots to eat the same meal before flying, thereby reducing the possibility of a common mode food-poisoning failure. But even with diversity, environmental challenges remain a threat to redundancy. As the examples of random failures indicate, multiple bird strikes or dense volcanic dust may cause both of an aircraft's engines to fail simultaneously.

Often, more complex forms of redundancy find use in systems deemed safety-critical, that is, systems whose failure is likely to lead to grave injury or loss of life. Redundancy requiring only one out of three or two out of four components to function further reduces the threat of system failure. Backup systems such as emergency generators find widespread use. Multiple backups may be called for, and if they are diverse enough to reduce common mode failures, so much the better. Diversity might require different manufacturers and even different principles of operation to further decrease the possibility of a common mode failure. At the cost of expense and complexity, redundant systems are a valuable addition to the necessity of reliable components in improving the safety of technological systems.

* * * * *

An insightful understanding of how things fail is paramount in the quest to produce technology with ever-increasing levels of safety. It is essential in eliminating early failures and in producing structures and machines that are robust in surviving the stresses of adverse environments and random events. It is vital in striving to slow aging and reduce wear. Where possible, design with redundancy and diversity further decreases the possibility of harmful failures. As engineers create technology, however, they cannot treat safety in isolation. They must also meet performance and cost goals, and often those requirements are in conflict with safety considerations. The next chapter examines the tradeoffs that must be made in technological design.

CHAPTER 3

Technology's Tradeoffs—the Engineers' Dilemma

THE MARINE PILOT and crew who took off from Elgin Air Force Base in Florida understood the importance of their mission. They were flying one of five prototypes of a revolutionary aircraft, one designed to perform maneuvers that weren't possible before. The Osprey was the first tilt-rotor aircraft. It could take off and land like a helicopter, but by tilting its rotors—or propellers if you prefer—through 90 degrees, it also flew like a conventional fixed-wing aircraft, cable of carrying troops at much higher speed and for much longer distances than was possible with conventional helicopters.

But Osprey engineering development was behind schedule and over budget, and to make matters worse, one of the prototypes had crashed, destroying the plane; the crew survived only though sheer luck. In an effort to assure continued funding for the development of the Bell-Boeing aircraft, the Osprey was on its way to demonstrate its

Figure 3.1: The V-22 Osprey Tilt-Rotor Aircraft, *US Navy*

unique performance capabilities and impress an audience of influential congressmen and other government VIPs.

On that summer day in 1992, the Osprey was to sweep in over the US Marine base at Quantico, Virginia, as the dignitaries watched, and then drop vertically to the tarmac in helicopter fashion for closer inspection by the admiring crowd. In his determination to complete his mission, however, the pilot ignored a warning light that came on in the cockpit shortly after takeoff. As the aircraft approached Washington, one of its two engines caught fire, and to the horror of the onlooking crowd, the Osprey plunged 500 feet into the Potomac River, killing all seven aboard.

The plane was grounded for 11 months while the crash was investigated. The cause was fluid that ignited after leaking from a gearbox. The earlier crash had resulted from a manufacturing error: incorrect wiring in the flight control system. Problems were corrected, additional flight testing performed, and despite some non-fatal mishaps, Congress continued to fund development. Then tragedy struck again—twice. In April 2000, marines were practicing night maneuvers in Arizona, but as one of the two Ospreys hovered before landing, it entered a vortex ring state—analogous to a stall in a conventional airplane. Losing its lift, the aircraft fell 2,000 feet, the crash killing all nineteen marines aboard. The following December, another Osprey crashed, killing 4 more marines. At that point, the military grounded the Osprey indefinitely.

The following years brought a thorough reanalysis of the Osprey's defects and major design changes to correct them. Extensive flight testing provided a better understanding of vortex rings, and engineers corrected the control system's shortcomings so that it would warn pilots if they were approaching the condition and display the appropriate action to avoid it. Investigators traced the cause of the second accident to the chafing of hydraulic lines, redesigned the engine hydraulic system, and eliminated the problem. Finally, in 2007, nearly 20 years after the first prototype took flight, the military declared the Osprey operational.

Since the Osprey achieved operational status, and experience has been gained, its reliability has improved to the point where then-Senator Obama was allowed to fly on the aircraft during a visit to Iraq. The Osprey's costs of manufacture and operation have slowly decreased as well. But the development of the revolutionary technology came at great cost. Project delays totaled well over a decade, and its costs were in excess of $16 billion. More tragically, the five Ospreys that crashed during development killed 36 flyers.

* * * * *

The history of the Osprey brings to mind a statement made by D. K. Price in his insightful book, *The Scientific Estate*:

When an engineer, following the safety regulations of the Coast Guard or the Federal Aviation Agency, translates the laws of physics into the specifications of a steamboat boiler or the design of a jet airliner, he is mixing science with a great many other considerations all relative to the purposes to be served. And it is always purposes in the plural—a series of compromises of various considerations, such as speed, safety, economy and so on.

This brief statement points to tradeoffs that engineers must make between three characteristics of the technology that they create: performance, cost, and safety. Substituting "the Osprey's performance requirements" for "speed," the quotation becomes as applicable to the Osprey as it is to steamboats, airliners or many other technologies. The design of technology invariably requires engineers to make many tradeoffs and compromises between what we here classify simply as performance, cost, and safety.

The tradeoffs between the three categories are more complex than they may first appear. Since each is multifaceted, tradeoffs inevitably appear within as well as between the three categories. They appear between performance criteria for speed, maneuverability and stability, and between development, manufacturing and operating costs. What is favored in design compromises varies markedly with the purpose that the technology is to serve. The Osprey was a military project, and the military by necessity places more emphasis on performance than it does on cost, or safety.

If your fighter plane, tank, or submarine cannot outperform that of the enemy, you might as well not build it, since it will be destroyed in battle. Cost and reliability are important, but not as important as superior performance. Conversely, cost and safety take higher priorities in commercial endeavors and consumer products. An airliner's price, fuel economy, and safety reputation count for more than whether its speed is marginally faster than its competitors. The engineering design process is full of such tradeoffs, from choosing between concepts—should the engines be jet or piston, air- or water-cooled?—to the most detailed of calculations—what diameter should the cooling fans have, and from what alloy should they be made?

Already there is talk of civilian use of the tilt-rotor aircraft for search and rescue operations, and for bringing relief to remote areas in the aftermath of floods and earthquakes. If this were to happen, the Osprey would be following in the path of many technological innovations introduced by the military that have eventually found widespread civilian use. Such transitions, however, require that the tradeoffs between enhanced performance, cost reductions and improved safety evolve from the military's requirements to those needed for commercial success.

The quest for performance often leads to implementing new technology—technology that is untried and its potential failure modes poorly understood before it is tested and deployed. It is no accident that many technologies—jet engines, sweptback wings, composite materials, satellite navigation and more—have been introduced by the military, and only much later—when their behavior and limitations have been more thoroughly explored—are they introduced for use in civilian endeavors. Fly-by-wire technology exemplifies such a transition from the military to the commercial sector.

* * * * *

Before fly-by-wire, both military and commercial aircraft employed cables guided by pulleys to mechanically transmit the pilot's commands from the cockpit to the aircraft's control surfaces on its wings and tail. The movement of these control surfaces channeled the airflow over the wings and tail, allowing the pilot to control an aircraft's climbs, turns and other maneuvers through the forces transmitted from the cockpit control stick and pedals to move the ailerons, stabilizers, and rudder. Such systems are reliable and durable, but the aircraft's design must be sufficiently stable so that the pilot's reflexes can easily compensate for wind gusts and other disturbances.

The Air Force and NASA initiated fly-by-wire research in the 1960s. Early on, the military's aeronautical engineers foresaw the advantages that would result from transmitting commands from the cockpit to control surfaces by electrical signals instead of mechanical cables. Both experimented by modifying existing fighter planes to incorporate fly-by-wire, with the first flights of the modified aircraft taking place in 1972. Initially, fly-by-wire was costly compared to traditional cable systems, and its reliability left much to be desired.

However, experiments showed that replacing cable and pulley systems with wires reduced both maintenance costs and weight. It also provided much more flexibility in the size and placement of the control surfaces, giving engineers more freedom to design for superior performance by changing the sizes and shapes of the wings and tail. By including a computer, the fly-by-wire system also can compensate for changes in flying characteristics when weight shifts due to depleting fuel, adding cargo or firing missiles. Equally important, when combined with computer control, the electrical signals allow pilots to fly and maneuver aircraft at supersonic speeds.

As experience was gained, costs came down and reliability improved. The military's quest for increased performance led it to be the first to adopt fly-by-wire. Subsequently, because of its necessity for effective supersonic flight, engineers brought fly-by-wire into the commercial realm with the development of the Concord supersonic airliner. More than a decade elapsed, however, before its adoption for other commercial aircraft.

In introducing the 757 and 767 airliners in the 1980s, Boeing relied on the tried-and-tested mechanical controls, even though digital fly-by-wire controls had by then become the nearly universal choice for military aircraft. The delays resulted from the need for the Federal Aviation Administration, its international counterparts, the aircraft manufacturers, and the airlines to be convinced that the safety of fly-by-wire was sufficient for use in commercial airliners. Concerns related to computer hardware failures, software glitches, and other potential failure modes had to be resolved before widespread commercial application of fly-by-wire could proceed.

NASA and the military had developed systems of redundant and backup computers, duplicate sensors, fault detection, and fault-tolerant systems so that—even if equipment failure or software faults occurred—the pilot would not lose control of the aircraft. Nevertheless, commercial aircraft manufacturers gave serious consideration to fly-by-wire only after years of military operation proved fly-by-wire's robust capabilities. Finally, in 1988, aircraft manufacturers, airlines and regulatory bodies gained sufficient confidence in fly-by-wire to employ it in commercial airliners, first in the Airbus 320 and shortly thereafter in the Boeing 777.

The extent of such design and testing precautions and the need to satisfy governmental safety agencies indicate why the design culture of commercial aviation was much slower than the military in adopting fly-by-wire technology. The five crashes and 36 fatalities attributable to the Osprey's design shortcomings during its development would have been the kiss of death for a commercial aircraft. Even a single crash—even a publicized near miss—could jeopardize the success of a commercial airliner.

The first crash of an Airbus 320 did indeed raise such issues. The crash took place at an air show during a demonstration flight of a chartered A320. The pilot executed a very low-altitude flyover in order to show the aircraft off to the crowd below. The plane smashed into the woods at the end of the runway, killing three of the 136 passengers. Since the A320 was just then being introduced into commercial service, new and innovative features of its design—particularly fly-by-wire—faced intense scrutiny following the accident.

Although some controversy remained, the authorities concluded that carelessness on the part of the pilot, rather than the aircraft's design, caused the accident. He was charged with involuntary manslaughter and served time in prison. Subsequently, the A320 became well established with the world's airlines, as did the Boeing 777, which suffered no fatal mishaps during development or early use.

* * * * *

Even within civilian technology, the ways in which engineers approach tradeoffs depends strongly on the nature of the product. Compare the design of a racecar, such as those that compete in the Indianapolis 500, and of a school bus. For the racecar, performance is everything: if you don't have a reasonable chance of winning the race, you might as well not build the car. Pushing the limits of technology increases the probability that the car will break down during the race, forcing it to withdraw, and may increase the risk to the driver as well. But if it doesn't break down, the potential for added speed will also increase its likelihood of winning. And racecar drivers are well aware that they are pushing their machines to the limit and engaging in a very dangerous sport, where driver error weighs more heavily than equipment failure in the risks they are taking.

Expense is secondary, and safety doesn't have as high a priority as it does in school bus design. Thus racecar designers are likely to implement daring new technology long before it has the proven track record required to use it in a family sedan or in a school bus. It is no accident that disc brakes, radial tires, turbocharged engines and other automotive innovations first appeared on racetracks.

In contrast, in school bus design, safety is paramount. Nothing must malfunction that might put the lives of children in danger. Moreover, the bus must be rugged because it doesn't operate on a well-designed racetrack under reasonable weather conditions. It must be capable of operating safely on potholed roads, in cold weather and hot, in rain and snow. Before new technology can be considered for incorporation into a school bus, it must be shown through extensive use and exhaustive testing to be rugged, durable, and exceedingly unlikely to fail catastrophically.

Conversely, the performance requirement for bus speed is minimal; it need only be able to travel at the highway speed limit for school buses. Cost, along with safety, becomes a far greater consideration than with a racecar because the bus must be affordable to local school districts or their contractors, whose financial resources are much more limited than those of the wealthy sponsors of auto racing teams.

Technological tradeoffs between performance, economy, and safety are inevitable, and the balance between them varies as much as the segments of the economy in which the structures and machines are put to use. Protection against fire, flooding, wind, and earthquakes compete with requirements for functionality, economy, and esthetics in the design of buildings as varied as homes, schools, factories, and hotels. Likewise, the tradeoffs for machinery differ greatly depending on its use. The tradeoffs between performance and safety, for example, are unique for recreation equipment. Consider, for example, all-terrain vehicles.

* * * * *

All-terrain vehicles—or ATVs—are four-wheeled gasoline-powered machines with soft oversized tires to facilitate off-road use. They are designed for a single rider; like a motorcycle, they have handle bar steering and a straddle seat. Some ATVs are used for utilitarian purposes, but in the US, most are purchased for recreation. And they have indeed become very popular. Since they first appeared in the 1980s, millions have been sold. Because many buyers seek the thrill of traveling off-road over rough terrain at high speeds, they demand larger ATVs with more powerful engines and greater maneuverability.

But those characteristics often come at the price of poorer stability; hence rollover accidents are a frequent cause of death and disability. It is a dangerous recreation, with some of the vehicles achieving speeds of nearly 50 miles per hour, and most states have no licensing requirements. The machines are particularly dangerous for novice users and for children whose parents often allow them to drive machines upon which they can barely reach the pedals. Hundreds of people in the US die in ATV accidents every year, and more than a quarter of a million suffer serious injuries. Worse yet, children suffer nearly a quarter of the deaths and injuries.

Designing ATVs that meet the thrill-seeking demands of those who buy them while providing some degree of safety presents conflicting challenges in making tradeoffs between speed, maneuverability, and stability. Pressure has grown for safer ATV designs to reduce the carnage from numerous rollover accidents that often crush the driver. A frequent question is why ATVs cannot have rollover protection, similar to safety features that have reduced the death rate from farm tractor rollovers. But the tradeoffs in farm tractor and ATV design are much different.

Farm tractors are heavier, driven at lower speeds and primarily on the rough but familiar terrain upon which a farmer must till his fields. Tractor safety is provided with a rollover bar imbedded in an enclosed cab, with a seat belt provided. In a rollover, the cab occupant is thus surrounded by a protective shell. But most thrill-seeking ATV enthusiasts would refuse to buy such a configuration. They would lose much of the pleasure of their sport, feeling that they were indoors instead of out, with loss of some visibility, and the rush of the wind. It would be similar to a lover of convertible automobiles having to drive with the top up.

A number of simpler roll bar devices, without an enclosed cab, have been proposed to protect ATV operators. However, with any safety device, care must be taken to ensure that it will lead to fewer rather than more injuries, and ATVs are a particularly challenging example. Even more than on a farm tractor, such devices tend to raise the ATVs center of gravity, making it less stable and more prone to rollover. Thus even if it lessens the severity of injuries in some accidents, it will tend to increase the number of rollovers.

Some rollbar configurations have an additional drawback; they would allow riders to hang on in precarious positions. But the tradeoff is even worse: for several devices that have been tested, the injuries from some types of accidents are more severe. Much depends on whether the driver is wearing a seat belt and helmet; often they are not. The device may provide head protection from the initial impact, but ATVs often do not just roll on their side, as a slower moving tractor tends to do; instead, they tumble several times. In these situations, the constraint system may actually result in more severe injuries because the likelihood increases that the driver will be pinned or crushed by the vehicle or undergo multiple impacts as it tumbles. The situation is very different from an automobile, or even a farm tractor equipped with a safety cab. In those vehicles, keeping the occupant within the passenger compartment nearly always increases their chances for surviving a rollover accident.

In some accidents, a safety device may cause injury. Thus, a measure frequently used in evaluating safety devices is the ratio of injuries caused to injuries prevented in comparison to the same accidents without the use of the safety device. International standards state that this ratio should be less than 7 percent and certainly no more than 12 percent. Seat belts and airbags in automobiles, for example, meet this standard. However, testing of several proposed safety devices to protect

ATV drivers have resulted in ratios that are much too high. Some rollover protection devices cause more injuries than they prevent. Such are the limits faced in the tradeoffs between safety and performance in a great deal of recreational and sporting equipment such as ATVs, snowmobiles, and jet-skis.

* * * * *

Automobiles occupy a huge place in most people's lives. They are the most common means of transportation for both work and pleasure, and mishaps in their use account for the largest cause of accidental death in the industrialized world. Predictably, the tradeoffs that engineers must make in automotive design strongly influence the nation's fuel consumption, accident statistics, driving habits and even the country's prevailing sense of style.

Conflicts between fuel efficiency and crashworthiness are pervasive in the tradeoffs that engineers must make. The most obvious way to increase an automobile's gas mileage is to reduce its weight. However, if no other changes are made to compensate for the reduced weight, the auto will be less safe in collisions with heavier vehicles. To see why this is so, you only need to make a simple calculation (simple for an engineer, that is) based on the law of physics, which states that momentum is conserved in the collision. Consider two vehicles of nearly identical design. One weighs twice as much as the other. They are both traveling at the same speed and collide head-on. Momentum is conserved in the collision, with the result that the impact felt by the riders in the lighter vehicle will be twice that experienced by those in the heavier one. Thus, there is a direct tradeoff between fuel efficiency and safety.

But as engineers consider more deeply the compromises they must make, they ferret out other ways to improve fuel efficiency without degrading safety, as well as ways to improve safety without reducing fuel efficiency. Innovation is often the key. But even then, other tradeoffs may come into play. Consider improvements in engine efficiency. In the 1970s, the dramatic increase in the price and even the lack of availability of gasoline resulted from the chaos in the Middle East. In attempts to lessen US dependence on foreign oil, Congress imposed 55-mile-per-hour speed limits, which both saved fuel and decreased the number and severity of accidents. The government also put fuel efficiency standards, known as CAFE (Corporate Average Fuel Economy), on automobiles. Through weight reduction and substantial improvements in engine efficiency, automotive designers were able to meet those standards. In fact, the innovations in engine design were so successful that the standards were exceeded.

As the price of oil came back down as the crisis eased, consumers paid less attention to gas mileage and safety and more to performance. Thus, the tradeoffs changed with market conditions. Drivers expected peppier performance in the form of higher acceleration. But this in turn made darting from one expressway lane to another to get around slower-moving vehicles more frequent. Such darting caused accidents and led to increased numbers of traffic fatalities.

Even more important, a loophole in the CAFE standard also led to the rise of the SUV, or sports utility vehicle. SUVs were built on a pickup truck chassis, and trucks were not required to meet the CAFE standards. This loophole allowed automakers to circumvent the CAFE standards and meet consumer demand for vehicles that were both larger and peppier. But the net result was to increase the safety of passengers in SUVs—a big selling point—relative to that of riders in smaller fuel-efficient vehicles. Again, using conservation of momentum, the larger the weight ratio of the larger vehicle to the smaller vehicle, the less severe the accident to riders in the heavier vehicle, and the more severe the accident to those in the lighter one.

Fortunately, over a period of years, other safety innovations have been introduced that lessen the negative impacts of peppier engines and larger vehicles. Antilock brakes lessen the probability of accidents through better control of the automobile, and airbags save lives by decreasing the impact on passengers when accidents do occur. But even with the new safety features, engineers still face the dilemma of tradeoffs.

* * * * *

Airbags represent a significant advance in automotive safety. But design engineers encounter tradeoffs and value judgments that tightly intertwine highly technical considerations with situations that are essentially political. Since their initiation, airbags have saved thousands of lives and reduced the severity of injuries in serious collisions. Their introduction, however, was not without problems, problems that stemmed from two important parameters that designers had to specify: the level of impact required to trigger airbag inflation, and the speed at which the bag inflates.

Raising or lowering the impact level that will trigger bag inflation results in mixed consequences. A lower triggering level is more likely to offer protection from all accidents. But in less-severe accidents, the airbag's inflation force sometimes causes worse injuries than those otherwise incurred from the accident. Conversely, a higher triggering threshold may allow too many injuries from impacts that are not strong enough to trigger the bag. Unfortunately, no triggering level exists at which the former problem is lessened without aggravating the latter. Engineers pick a level that attempts to minimize the total number of injuries. But their decision will not prevent lawsuits from those injured in accidents too mild to trigger airbag inflation and those harmed by the deployment of the airbag itself.

The force of high-speed bag inflation has killed a number of children and small adults. Designers had set the inflation speed high enough to prevent injury to an adult who was not wearing a seat belt. If the engineers were allowed to assume that adults would wear seat belts, then they could have reduced the inflation speed, thereby also reducing the chance of injury or death to smaller passengers. Thus, the designer faces a social question: Should the inflation speed be chosen to optimize protection for the large numbers of teenagers and adults who refuse to buckle up, or to provide maximum protection to the smaller numbers of child passengers?

Over time, this dilemma has been all but eliminated: newer airbag systems are complex. They inflate with a speed that increases with the severity of the collision, and they include sensors to determine the passenger's weight and whether he or she has a seat belt fastened. With this information, electronic controls adjust the airbag inflation speed accordingly. But with these more complex systems, the frequency of bag failures may increase unless engineers adequately address possible sensor and adjustment mechanism malfunctions to ensure that the airbag systems maintain their reliability. Once again, detailed design tradeoffs are in fact not only technical but political and social in nature as well.

Airbags, of course, are not unique in raising complicated tradeoffs in determining adequate levels of safety. Questions arise in weighing the risks from one class of failure with another: specifying a lower tire pressure, for example, increases traction on slippery pavements, but it also increases wear and the likelihood of catastrophic failure in high-speed, long-distance travel, particularly in hot weather. The effectiveness of airbags, better tires, or other safety improvements must also be weighed against offsetting the propensity of drivers to behave less cautiously as they gain confidence in the safety of their vehicles, a topic to which we shall return.

Other situations pit the risks of different stakeholders against one another: safety guards on food-processing machinery reduce the frequency of occupational injuries. But the guards also make cleaning more difficult, thus increasing the risk of food contamination. Similarly, specifying the delay between the time when railroad crossing gates go down and the time when the locomotive passes over the intersection requires a tradeoff. Lengthening that delay will encourage more impatient motorists to circumvent the gate and result inevitably in disastrous collisions; reducing the time may leave slower-moving or impaired vehicles stranded on the tracks in the path of the oncoming train.

* * * * *

The myriad decisions made in a design—of everything from microprocessors to supertankers—mix technical, social, economic, and political considerations in complex combinations. How society deals with technological risk and determines how safe the resulting technology is involves design organizations, regulatory bodies, insurance underwriters, and those who purchase and use the end products. Such groups' interactions are far from simple, and the tradeoffs they must make often seem at odds with the simple ideal of minimizing the total number of deaths or the amount of suffering that accompanies the benefits of what the technology might bring.

The balancing of design concerns differs between civilian and military ventures, as well as between cultures, industries, and corporations. As engineers weigh the importance of performance, cost, and safety, the balances they strike must be as varied as the clients and customers they seek to serve and as the regulations to which they must conform.

CHAPTER 4

Finding Faults—Exposing Hidden Hazards

The spectacular collapse of the eight-lane I-35W Mississippi River bridge in Minneapolis in August 2007 caused 13 to fall to their deaths and more than a hundred injuries. The disaster brought nation-wide media attention in the weeks that followed, and many assumed that structural deterioration or faulty maintenance was responsible. Further investigation, however, revealed that the cause was a design flaw, built into the bridge when it was constructed some 40 years earlier. A steel plate was not thick enough to securely connect the junction of girders that it held together. Designed too thin to begin with, the plate underwent additional stress when modifications—the addition of a median barrier—increased the weight of the bridge. The faulty design caused cracks to form prematurely and grow, leading to the lethal collapse.

In the more than 40 years since that design flaw was built into the Minneapolis Bridge, the methods for finding and eliminating

Figure 4.1: Collapsed I-35W Mississippi River Bridge,
Federal Emergency Management Administration

flaws in design have improved. More powerful computer simulations calculate stresses in much greater detail and model more accurately the forces to which a structure is likely to be subjected. But the processes for assuring that flaws do not survive the processes of design and construction are not perfect; it is a never-ending task requiring ever-present vigilance. Concern for safety must be present at the earliest stages of design, as engineers study the desires and needs of likely customers, trying to understand how they will interact with new or refined products or services, and also how their use—or misuse—might lead to hazardous outcomes.

In verifying their success in the performance and cost of the technology they create, technologists must also continue their quest for safety. Engineers scour their concepts for hidden hazards by continually asking "what if?" Prodding their imaginations to uncover potential hazards, they frequently use study aids: they create fault tree diagrams, which logically display the many ways in which the technology could conceivably fail. They construct lists of potential failure modes and examine the effects that each might have. They take advantage of the ever-advancing state of computer software to simulate the technology's behavior, allowing many problems to be solved and hazards eliminated before construction or production is begun.

Engineers must go beyond experience, theory, calculation and simulation to ensure that what they create will meet the performance and cost requirements that have been set. Moreover, they must assure that the technology is reliable and that it does not create hazards for those who use it, or for the wider public. The methods for reducing technology's risks vary as widely as the products and systems to which they are applied. They differ greatly between automobiles, appliances, medical devices and other mass-produced products, and one-of-a-kind dams, bridges, and buildings. They differ between computer software, one-of-a-kind ships, and continent-spanning gas pipelines, electrical grids, and air traffic control systems.

Bridges, such as the one that collapsed in Minneapolis, offer an example of how unique buildings, bridges, or other structures, must undergo scrutiny to assure that what is built doesn't contain hidden flaws that may lead to future disaster. Laboratory testing of components, particularly those that are new or uncommon—for which long experience in their use has not already been gained—in an essential part of the design process. Giant strength-testing machines, often located at universities or other research institutes, stretch, bend, and twist beams, joints and other components to assure that they can resist the loads and punishment beyond what the bridge is expected to experience. But component testing is not enough. Once the bridge is built, engineers must confirm with proof tests that what they have created is as strong as they designed it to be.

The first proof tests to be performed are of a diagnostic nature, to assure that the engineers' earlier calculations are confirmed in reality. Loaded with the number of trucks or other vehicles that the structure is expected to carry, engineers place strain gages or other instruments throughout the structure and measure the strain and deflections that actually exist, then compare them to those predicted by their design calculations. Usually, the quantities will be smaller than those calculated, indicating that the design calculations were "conservative," that they overestimated the stresses in the bridge. This, of course, doesn't preclude the possibility that the bridge will not be as strong as the design calculations predicted.

Structures are designed with safety factors to assure that loads much larger than those anticipated in normal use can be sustained without damage to the structure. To verify that the bridge can indeed sustain such dead weight, it is heavily loaded, for example, by crowding it with trucks packed to capacity with concrete, or by deliberately placing heavy stationary weights on it. Even the Army's battle tanks have been used because they are able to concentrate very heavy weight on small stretches of a bridge. Thorough inspection then takes place to assure that the strain of the bridge does not exceed safe values.

The locations where bridges, buildings or other structures are built have strong implications for the stress they must withstand, and engineers must use means other than proof testing to minimize the risks of catastrophic failures. Depending on the bridge location, protection against corrosion from salt water, freeze-thaw cycles, high tides and flooding, hurricane-strength winds, collisions from barges or ships, and—perhaps most of all—earthquakes must be central to design considerations. Engineers cannot eliminate the risks from such disasters—natural or man-made—by employing only proof testing methods. The design challenges imposed by improbable but potentially catastrophic events will be a recurring theme in later chapters.

* * * * *

Bridge failures, building collapses, sinking ships, airliner crashes, or other spectacular accidents elicit a great deal of attention from the media and the public. Fatal design flaws that escape detection in mass-produced products, however, tend to result in accidents—most having only a single victim—that initially may attract much less publicity than disasters in which dozens die. But the numbers add up, and such flaws may cause far more fatalities than single disastrous events, such as the Mississippi River bridge collapse.

Calculations and computer simulations, the study of the strengths and weaknesses of earlier designs, and the testing of components are invaluable in predicting how a technological creation will behave once it is put into use. But in confirming that their creations actually perform as predicted and that hidden safety hazards do not remain when the technology is put to use, the methods used for mass-produced products differ significantly from those employed for one-of-a-kind technology. For mass-produced products, testing of prototypes plays a central role in this process. While the testing methods vary as much as the nature of the product, automobiles—one of the more complex products produced by the millions—offer a pertinent example.

Automobiles are a mature technology. The extent to which a new design closely parallels earlier models provides some confidence that things will work as planned. But new, innovative, and complex components need to be tested early on. Engineers build models and mockups, run engines, and test components. They come as close as possible to approximating real-world environments. They do this by installing newly designed brake, steering or other mechanisms in existing automobiles and test-driving them over rough roads, in severe weather and more. They use whatever means they can devise to ensure that the more innovative features of the new design will not perform poorly, or worse, cause an unanticipated safety hazard. Next, they build costly operational prototypes to road test their newly designed vehicle as a whole.

Prototypes must be available for testing well before mass production facilities capable of producing automobiles at rates of more than a thousand per day are ready to operate. Typically, automotive design verification and validation requires 300–400 prototypes. Thus, the corporation builds a small pilot plant where highly skilled technicians and trade workers craft prototypes one at a time, without the aid of robots or assembly line organization. Some prototype parts are off-the-shelf, mass-produced items; the standard nuts, bolts, batteries, mufflers, gaskets and so on. But many are unique to the design, produced by shops in small numbers specifically for the prototypes. The production rate of such a pilot operation is likely to be closer to one per day than the one per minute target of a typical assembly line. This translates into a large cost per vehicle; the cost of constructing each handcrafted prototype is likely to exceed ten times the selling price of the mass-produced vehicle.

Carefully crafted and thoroughly checked as they are, the prototypes are not subject to the infant mortality failures that stem from imperfections in mass production procedures. Instead, their purpose is to serve to flesh out and correct design flaws that could cause random and wear failures. The autos are run over test tracks with all sorts of punishing surfaces and in all sorts of weather. They are sprayed with salt water, covered with ice, driven through floods, and more. In North America, some operate in Canadian winters and others in Mojave Desert summers to assure that they can survive the extremes of cold and heat. This and more is required to assess the limits of environmental stress that the vehicle can withstand. More than 100 prototypes are fitted with instrumented crash dummies, and smashed in front, side, and rear collisions to verify the auto's ability to meet crashworthiness standards and regulations.

Durability tests to predict the longevity of the design and to uncover safety hazards that may only appear after years of operation present another set of challenges. Prototypes may be driven 50,000 miles or more through mud and salt water, over gravel, potholes, cobblestones, broken glass, and other test track obstacles to simulate wear and tear well beyond what most vehicles will experience in 100,000 or more miles of operation. Engines are started and stopped more times than anticipated in an auto's life, and wear is monitored to set and refine maintenance requirements.

Progressing from pilot plant prototyping to assembly line production, engineers must deal with the threat of infant mortality failures. They must ensure that defective automobiles don't result from poorly adjusted machinery, inattentive line workers, or a supplier's defective parts. Just as design flaws are the design engineers' nemesis, manufacturing defects are

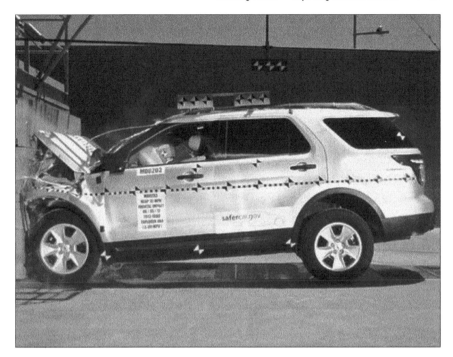

Figure 4.2: NHTSA Crash Test,
National Highway Transportation Safety Administration

the production engineers' fear. They must assure that high-speed parts manufacture and assembly operations, carried out by hourly workers and robots, faithfully replicate the automobile that emerged from the careful crafting of the prototype pilot plant.

The engineers face an added challenge: Exhaustive test drives to verify that each automobile is produced as it was designed would be far too costly. High rates of production allow sampling of only a small fraction of the finished automobiles to verify that their performance meets all the standards for which they were designed. Thus, defects must be detected and eliminated early, before the autos roll off the assembly line. Such are the challenges of producing large numbers of unflawed products that are both affordable and safe.

All such efforts are made to prevent the designers' nightmare that a serious design flaw or manufacturing defect will escape detection and correction, only to become apparent after the car is on the market and millions have been sold. Of those millions, thousands will be involved in accidents, many of them serious, and each with a cause. Most are caused by driver error, often abetted by intoxication, drowsiness, or distraction. But as incidents are reported and complaints come in, events that seem to point to a defect in the car itself are of greatest concern.

Corporate officials and engineers alike hope that none point to a serious design flaw, but rather that the causes are random, resulting from dangerous driving habits, unfavorable circumstances, or faulty maintenance. Even then, lawsuits are likely to result, and the manufacturers are required to defend their product's safety. But far scarier are the driver complaints and accident reports that seem to involve a particular problem with the automobile, whether it relates to brakes, tires, steering, or stability. Certainly, at the first signs of trouble, engineers will reexamine relevant systems and components to determine whether there is a design flaw that they missed in their design and testing, and if so to remedy it.

Ethical dilemmas begin to arise. At what point has enough evidence been gathered that a car should be recalled? If subsequent investigations show that the accidents were not caused by faulty design or manufacture, a recall may do undeserved harm to the automaker's reputation, damaging sales and reducing profits—harm that the engineers, and especially the company's higher management, desperately want to avoid. But if it turns out that a design flaw is to blame, the company needs to get ahead of the problem, to recall the cars and fix the design problem, before the National Highway Traffic Safety Administration or another regulatory body orders them to make the recall; this would be a much larger blow to the company's reputation. Far more important, the longer the problem is left to fester, the more accidents will ensue, bringing more deaths, injuries, and lawsuits.

If a recall is issued, then the autos' owners expect a prompt remedy to the problem. They don't want interminable delays, with their car out of service at the dealers, and they want assurance that the safety hazard has been eliminated. But this may be problematical. The engineers must be able to diagnose the problem's precise cause, for only then can they redesign a part or system to eliminate the hazard. Both diagnosis and redesign may require time. The uncertainty is reduced only as accidents are investigated and the causes are determined definitively.

Regulators must be convinced that the remedy is adequate. Then contracts must be made, and repair parts must be manufactured and delivered to the dealers, along with instructions for their proper installation. Moreover, the manufacturer must make sure that the dealers' service departments are competent enough to make the repair properly. Otherwise, there is the danger that faulty recall repairs will lead to yet more accidents. Each step takes time, and a mad scramble is likely to ensue. Delays will aggravate customer irritation, and dealers' inventories will remain unsold—as potential customers migrate to competing brands—until the repairs can be made.

* * * * *

The problems that arose in uncontrolled acceleration incidents with Toyota in 2009–2012 illustrate the complexities of dealing with design flaws that are not identified until millions of automobiles have been sold. The problem first surfaced in 2009, when a number of reports mentioned vehicles experiencing unintended acceleration.

Among the most dramatic was an August 2009 accident in California. A floor mat caused the accelerator pedal to stick, and the driver—an off-duty highway patrol officer driving a Lexus—tried desperately to bring the vehicle under control. As the auto sped though busy intersections, accelerating to speeds of more than 100 mph, one of the car's occupants frantically asserted in a 911 call that the accelerator was stuck and the brakes were not responding. The last words that the operator heard were "hold on and pray" followed by a scream. The Lexus clipped another vehicle, crashed though a fence, careened down an embankment, and then bust into flames as it came to rest in a riverbed below. The vehicle's four occupants all died of their injuries.

Understandably, the accident drew a great deal of media coverage and brought the gas-pedal problem to the forefront of the public's attention. Investigating this and other accidents involving Toyotas, the

National Highway Traffic Safety Administration and others brought about a widespread 2009 recall to eliminate gas-pedal entrapment caused by out-of-place floor mats interfering with pedal movement.

But the uncontrolled acceleration issue turned out to be yet more complicated than solving the floor mat entrapment problem. A second recall took place in 2010 after reports came in of uncontrolled acceleration crashes that could not have resulted from gas-pedal entrapment because the floor mats had been removed in some cases. Following the recall announcements, media attention caused a sharp increase in the number of problems reported and in the number of alleged victims.

A possible cause for some accidents was thought to be mechanical sticking of the gas pedal, likely resulting from wear and aging. Engineers redesigned the pedal assembly and eliminated that problem. However, replacement parts could not be supplied to the dealers fast enough, and Toyota had to suspend sales of the models subjected to recall. In the US, some owners reported accelerator pedal problems even after repairs had been made, but in these cases investigations found the cause was improper repairs.

Far greater public attention became focused on modern vehicles' electronic throttle control systems, called drive-by-wire, in which the gas pedal is linked to the engine electronically rather than through the mechanical cables used in earlier years. Much was at stake; many lawsuits and some safety advocates contended that software glitches were the cause of the unintended acceleration. Following prolonged public debate, the issue was settled only after the National Highway Traffic Safety Administration (NHTSA), working with NASA scientists, conducted a 10-month investigation that concluded that the electronic throttle control was not responsible for the problems. They examined many thousands of lines of code in nine vehicles involved in sudden-acceleration incidents and found no problems.

NHTSA experts also examined the black boxes from 58 vehicles involved in sudden-acceleration incidents and found that, in more than 75 percent of the cases, the brakes had either not been applied at all or had been applied just before the crash. These findings provided strong evidence that drivers had been pushing on the wrong pedal.

The final report listed three causes: pedals becoming trapped on floor mats, pedals that can become stuck as a result of wear and aging, and—most frequently—driver error.

Before all the issues were resolved, NHTSA forced Toyota to pay $49 million in fines—the maximum allowed by law—for delaying too long the recalls related to sudden acceleration. Worldwide, Toyota recalled more than 14 million vehicles, largely as a result of the floor-mat problems and the accelerator pedal design defect. Nearly 40 deaths were thought to result from the gas-pedal problems.

* * * * *

Each mass-produced product has its own characteristics and market. These require the methods used to reduce its risks to vary greatly between products. Like automobiles, medical devices are mass-produced, and many thousands of pacemakers, defibrillators, artificial hips and knees are implanted in patients each year. Efforts to uncover and eliminate design flaws and ensure the safety of implanted devices confront challenges that are quite different from those encountered in automotive prototype testing.

If a potentially dangerous design flaw is found, the manufacturer can recall an auto to get it fixed; the owners are faced only with the inconvenience and expense of getting it to the dealer to have the repair done. But if that design flaw is in a pacemaker or an artificial knee joint, major surgery is required to replace it with a more reliable design. Such surgery, however, has is own risks, and these may be great for patients who may be frail and have other major medical problems. The risks of the surgery must then be carefully weighed against the chance that the device will fail, causing loss of quality of life, and too frequently death.

The approaches to prototype testing relevant to automobiles, aircraft, computers, and consumer products are not appropriate for devices that will be implanted into human beings. Instead, more reliance must be placed on laboratory testing—referred to as *in vitro*—to assure that the device materials are compatible with the environment they will encounter when implanted in human tissue. They must not leach away to cause toxic or allergic reactions, even though the corrosive salty environments found in our bodies aggravate the devices' deterioration and failure. Computer simulations to assess mechanical stress levels and fluid interactions to which the devices will be subjected augment laboratory testing using chemical and biological agents. In some cases, animal testing—using mice, hamsters, or rabbits—may also be necessary, since living tissue may better predict how the devices will behave when implanted in humans.

The closest analogy to prototype testing is the clinical trials that devices must undergo before they can be granted approval by the Food and Drug Administration for widespread use. Professional drivers of prototype automobiles and the test pilots who fly newly designed aircraft certainly assume some of the risk encountered in introducing new technology in order to greatly reduce the risks to the public. Likewise, risks are taken by the volunteers who take part in the clinical trials. Although they are carefully monitored for signs of trouble, they are exposing themselves to whatever design flaws may have gone undetected and uncorrected in the laboratory and animal testing before surgeons implant the devices in their bodies. Of course, they needed a device to improve their quality of life or increase their longevity. But for the prospect that the newly designed device will relieve their malady, they are willing to assume the risk that a design flaw may cause the device to fail, with all the adverse consequences that failure could bring.

As with other technologies, scrutiny cannot end with a regulatory body's approval of a device for general use. Clinical trials are relatively short-lived, and longer-term deterioration must be monitored. Likewise, the patients in whom the device is installed have a much broader set of characteristics than those of the limited number of people in the clinical trial: differing states of health and medical problems with which the device may interact, and differing environments in which the patients live.

A challenge common to innovative technology—but particularly pronounced in the case of implanted medical devices—relates to seemingly small changes in design. In the first few years following the introduction of a new medical device, designers learn from clinical trials and from the use by the wider population how to improve the performance, economy and/or reliability of the device though numerous small changes. Requiring the modified device to go through the same testing procedures and clinical trials as the original design would likely make testing prohibitively expensive for the small advantages gained; therefore, the project likely would be dropped. Even if it weren't, the clinical trials would greatly delay the public's access to the improved devices and the heath benefits that patients would likely derive from it.

But conversely, what if the improved design also introduced a harmful design flaw that would not be caught without clinical testing? The danger then is in introducing that flaw into many thousands of devices implanted in patients. Risks from both direct failures and from the many surgical operations to replace the defective devices would then be incurred.

Government regulatory bodies must take such considerations into account and draw a line—or lines—beyond which the design changes are so insignificant that much of the testing can be bypassed. For medical devices, the Food and Drug Administration employs a streamlined procedure that allows clinical testing to be bypassed if the modified device has "substantial equivalence" to the earlier device. But knowing where to draw that line is difficult, and mistakes are possible. One of the most publicized mistakes relates to the ASR XL artificial hip introduced in 2005 by DePuy.

The modified replacement hip launched by DePuy differed from the earlier model in that the ball at the top of the femur and the socket liner inside the pelvis—as shown in Fig. 4.3—were made of the same chrome-cobalt metal; in the earlier model, the ball was metal, but the socket liner was plastic. The main reason that metal-on-plastic hips failed stemmed from wearing down of the plastic socket. Laboratory tests seemed to confirm that metal-on-metal hips would work well and last much longer, eliminating the need for surgery to replace worn sockets. As the implantation of all-metal hips became widespread,

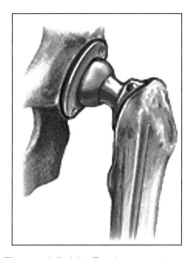

Figure 4.3: Hip Replacement,
National Institute of Health

however, serious problems began to appear, most frequently in the form of debilitating pain, but there were other symptoms as well. Slowly, as investigators gathered statistics, the extent of the problem became known and the mechanism understood. As the ball and socket rub together during normal use, metal debris—in the form of microscopic particles—is shed, damaging the surrounding tissue and even entering the bloodstream. By the time this evidence could be gathered and a recall ordered, nearly 100,000 hips had been implanted. Not all of the hips are likely to be replaced, but many surgical operations are taking place—and lawsuits filed—as a result of the misjudgments in predicting the consequences of a design change that brought disaster rather than relief. Not only did the device have to be withdrawn from the market, but the malfunctioning hip replacements are likely to cost the company billions of dollars to settle law suits stemming from the defective design.

* * * * *

Infrastructure—whether it is an electrical grid, rail systems, or water supply or sewage removal—presents yet another set of challenges; it must remain in operation while it is being expanded, upgraded, or replaced. This poses unique safety implications for the engineers who implement such projects. The greater the change from the existing system, the greater the challenges are. If the infrastructure is not simply being expanded or upgraded, but being replaced by a new system with the promise of much improved performance, the challenges are greatest of all. It is somewhat akin to standing on one foot to install the new system while trying to keep the old one operating. Such was the case, for example, in converting the nation from traditional analog television broadcasting to the digital high-definition system. But that transition had few safety issues.

In contrast, the Federal Aviation Administration's current program to replace the aging air traffic control system with a twenty-first century system must do so without even a single aircraft accident resulting from problems in the transition. It must meet the dual challenges of simultaneously implementing the new system while maintaining safe operation of the current air traffic control system until the new one is fully installed.

NextGen—short for Next Generation Air Transport System, the Federal Aviation Administration's revolutionary new satellite-based system—is planned to replace the aging air traffic control system based on radar and voice radio communications. While the present system has served the nation well—as witnessed by its ability to handle more than ten million flights each year, often without even a single crash—it nevertheless has become increasingly outdated.

The radar-voice system has a number of constraints. Radar's range is limited, and controllers must pass control from one region to another, introducing chances for miscommunication. Mountains, oceans and remote landmasses lead to blind zones which aircraft must pass through. The accuracy with which radar can locate an aircraft is limited, and it provides no information on the planes' altitudes.

Continuing to operate with the present system in the future will eventually bring gridlock to the sky and to many airports as the number of commercial flights continues to increase. More important, since the skies have become more crowded, handling the traffic with the radar-based system becomes more difficult; one consequence is the growing number of midair near-misses between aircraft and collisions of taxiing aircraft at busy airports.

To deal with these increasingly severe limitations, the Federal Aviation Administration is implementing the multibillion-dollar NextGen project. This system replaces ground-based radar and voice communication with a highly automated satellite-based system employing GPS—the Global Positioning System—to locate and track aircraft. Its increased capabilities should allow more planes, helicopters and even drones to occupy crowded airspaces, but with added margins of safety. The FAA predicts that NextGen will be able to shorten routes, thereby saving time and fuel and reducing the emissions of greenhouse gases; it should also avoid delays by eliminating the need for airport "stacking" as planes wait for their turn to land.

The FAA predicts that NextGen will increase the safety of flying. It will improve air traffic controllers' capabilities in identifying and guiding aircraft because the GPS system will be much faster than the call-and-response system that has been used in conjunction with radar for decades. In addition, GPS equipment will accurately locate the positions of aircraft and show their altitude, speed and direction of travel. It is designed to operate seamlessly across mountains and oceans, relieving air traffic controllers from the error-prone handoffs between geographic regions.

Cockpits equipped with the system will allow pilots to see on their displays the locations and altitudes of similarly equipped aircraft that are flying within hundreds of miles and to view their flight plans as well. It will also provide pilots and controllers with up-to-date weather data, allowing for rapid adjustments of flight plans, and it will warn oncoming aircraft of potential collisions and prescribe coordinated evasive action to both cockpits.

But bringing the new system into being is a challenge; if it is not carried out meticulously, it could create some safety hazards even though its completion would make air travel even safer than it is today. Bringing in a new system while continuing to operate the old one is never a straightforward task. It must be phased in carefully, using designated airports and regions as test beds for equipment and procedures to work out any unanticipated safety issues before the system is applied more widely.

Projects involving millions of lines of computer code inevitably turn up omissions and errors that must be identified and eliminated without endangering the public, and NextGen is no exception. Indeed, computer simulations and testing at initial operating sites has already identified more than 200 software-related problems, such as data processing failures, errors in aircraft identification, and hand-off problems between controllers. Moreover, security and defense against hackers are continuous concerns. The GPS aboard aircraft will continuously transmit radio signal packets identifying location, altitude, flight number and the like. These must be securely encrypted to prevent criminals, terrorists, or others from gaining access to sensitive information and using it to bring havoc to air traffic control.

The FAA and the airlines must install NextGen equipment in control towers and aircraft, and air traffic controllers and pilots alike must be trained to effectively utilize the highly automated new system to manage their workloads, while at the same time retaining their diagnostic skills and ability to operate the existing system without errors. The transition must be brought about in such a manner that the controllers and pilots are not confused by the presence of both old and new systems in towers and cockpits. All this must be accomplished while the old system is kept in operation 24 hours a day, 7 days a week, without allowing a single mistake that leads to loss of an aircraft. Pilots need to understand how they should respond to the new alert systems because the increased safety and efficiency of the GPS-based system allows aircraft to operate closer together.

* * * * *

Bridges, automobiles, medical devices, air traffic control systems, or any other of the diverse forms of technology that permeate industrial society: each brings unique engineering challenges to assure that flaws in design and implementation do not go unrecognized to create hidden hazards that lead to undue risk to the public. The search for such flaws must extend further than attempting to locate shortcomings inherent to the technology itself, whether they be in hardware, software, or systems design. The search for flaws must extend to minimizing the likelihood that the people who build, install, operate, and maintain the technology will make errors that may lead to dreadful consequences. Technology's creators must strive to make it difficult for drivers to step on the wrong pedal, to facilitate surgeons' error-free implantation of devices, and to eliminate ambiguities that might lead to pilot or air traffic controller error. The following chapter examines the interactions of humans with their technology more closely and seek to understand how the incidence of human error that contributes to all too many accidents can be reduced or eliminated.

CHAPTER 5

To Err is Human

In the summer of 2006, Denise Melanson of Rainbow Lake, Alberta, a 43-year-old teacher and mother of two, was diagnosed with an advanced but treatable cancer of the nasal passage. Instead of remaining in the hospital for chemotherapy, she was given an electronic infusion pump containing a four-day dose of the chemotherapy drug fluorouracil. She was to administer it to herself at home.

The pump, however, had been mistakenly programmed to dispense the powerful drug over four hours instead of four days! Upon realizing that the medication had run out long before the prescribed four days, she immediately returned to the hospital. The error was then obvious, but it was too late; she died three weeks later from the toxicity of the overdose.

How could this have happened? Investigation led to the conclusion that a nurse had incorrectly calculated the dose that she programmed into the machine. The nurse's error resulted from trying to convert a dose rate based on milligrams over four days to one based on milliliters per hour. She had performed the calculation with pen and paper because no calculator was readily available. A second nurse checked the calculation, but without a calculator, she committed the same error, and so the dosage was approved.

The programming of the infusion pump was error-prone. It required a complex procedure of scrolling though options with no feedback or other safeguards provided. In a controlled test, the staff at another institution recreated the same set of circumstance and attempted to program the infusion pump; they made the same error. At least seven patients have died as a result of similar errors in other North American cancer centers.

Such deaths can be attributed to human error; in this particular incident, it was committed by well-trained nurses. But no doubt those nurses were multitasking, often pulled away from other duties to program the medical devices. Arguably, the tragedy was more the fault of poorly designed technology that required overworked hospital staff in stressful situations to quickly and accurately convert milligrams over four days to milliliters per hour! Where lives are at stake, improved software and more user-friendly displays should eliminate the need for such on-the-fly calculations, provide clear indications of the results, and incorporate safeguards. One would hope that improved software and displays on the pump would clarify the unit conversions and obviate the necessity for nurses to perform them by hand.

Even more frequent mistakes involving the amount or rate of dispensing medications have too often led to deadly consequences. For example, infusion pumps are widely used to deliver measured amounts of medication to the bloodstream though IV injection systems over a specified length of time. The dosage, rate of delivery, and other data must be properly entered onto the pump's display screen for the medication to be correctly delivered. However, confusing instructions or poor design of the display screen may not make clear how the patient data is to be entered; units of measure (are they, for example, in grams or ounces?) may not be sufficiently clear, leading to incorrect dosages.

Not only equipment for handling powerful drugs, but also other medical procedures requiring medical personnel to program complex devices, frequently result in human errors that lead to fatal results. Overdoses of radiation delivered through X-ray machines, CAT scans and, in particular, the large doses employed for cancer therapy can be deadly if the technicians who operate the complex devices make mistakes.

New technology, such as linear accelerators, allows physicians to target tumors more accurately and deliver doses of radiation precisely, greatly increasing cancer patents' chances of survival. Yet, as studies such as that performed by *The New York Times* indicate, hundreds of patient deaths and injuries result from radiation overdoses stemming from the use of such complex medical technology. Most often, a combination of factors including inadequate training, lack of familiarity with newly installed devices, excessive workloads, and operator attention lapses contribute heavily to these deadly mistakes.

But case histories also indicate that the frequency of treatment errors could be substantially reduced by more thoughtful designs of computer displays and other interfaces, and through improved software that is unambiguous and straightforward to use. In many situations, a better design of medical devices in respect to human factors could alert operators that an error is being committed. More carefully designed interlocks would assure that procedures are sequenced properly and block actions that violate preprogramed safety limits.

Even where electronic devices are not involved, poor design of human-technology interfaces can lead to fatal results. Consider, for example, the complex web of tubing that invariably connects critically ill patients to various medical devices. In such situations, deadly mistakes may result from ambiguous connector design. For example, often the patient is hooked to IV tubing to insert medication directly into the blood stream. If the patient cannot swallow, a second system of tubing will simultaneously feed nourishment into his stomach. But the bags for the medication and feeding solutions must be changed and refilled.

Multiple deaths have resulted when the feeding solution from an external feeding tube has been inadvertently introduced into the IV tubing, causing it to enter the bloodstream. Such misfortunes are possible when adapters for connecting nourishment bags to feeding tubes can also fit into the IV system. Such mistakes could be eradicated with more careful design of the tubing connectors so that those for the IV and those for the feeding tubes are incompatible with one another. But achieving such a goal would require equipment manufacturers to adhere to a uniform set of standards; only such uniformity would eliminate mistakes, since hospitals invariably combine equipment purchased from multiple manufacturers.

The millions of tasks performed each year utilizing medical technologies provide salient examples of the dangers caused by the use of poorly designed equipment, even in seemingly routine situations. Each year, human errors committed during medical procedures are estimated to cause tens of thousands of deaths. Physicians, nurses, technicians and others are prone to such mistakes, since they often work under high stress levels, making rapid decisions while following complex procedures and dealing with multiple patients and medical devices.

We may call these "human errors" because rarely are they caused solely by device failure. Additional training may not help; most often, the errors do not result from lack of knowledge, but rather from information overload, momentary lapses of attention, distraction, or fatigue. However, more careful examination of the situations under which such errors have occurred illustrates that more attention to human factors in device design makes mistakes more difficult to commit and thus greatly increase the safety of medical procedures. Substantial progress has been made in improving human interfaces with medical technology, but guarding against the introduction of error-prone devices is a never-ending task.

* * * * *

The root causes of many accidents across the wide spectrum of technological endeavors are attributable to human error, and often that error is exacerbated by poor design of human-machine interfaces.

Pilot error is frequently cited in investigations of aircraft crashes, just as operator errors frequently play critical roles in figuring out what happened in train wrecks, truck rollovers, crane collapses, refinery explosions and the like. Even if an accident's cause is an equipment failure, extreme weather, or other phenomenon outside of human control, decisive actions by those who operate the technology—the drivers, captains, pilots, control room personnel and others—frequently can mitigate or exacerbate an accident's consequences. Consequences often depend on the decisions that operators make, decisions that they must make while under a great deal of stress, where an inappropriate action may make the situation much worse.

Training and experience are of course vital in reducing the errors that operators make. But poor design of the interfaces with which humans and technology interact, interfaces that do not give operators the information they need or that are confusing, is a frequent cause of human errors. Life-saving increases in safety often result from carefully examining the design of technology and focusing attention on how humans interact with it.

Design that is less prone to cause human error first began to gain attention during World War II: investigations made clear that unacceptable numbers of aircraft crashed because pilots too frequently would grab the wrong lever or engage the wrong switch on their controls due to the stressful conditions. This led to efforts to redesign cockpits to reduce the chances of pilots making such mistakes. These efforts led to the field of human factors engineering, much of which is devoted to improving human-technology interfaces so that operator errors are less likely to happen.

Human-technology interfaces take many forms: controls on everything from automobile dashboards, ships' bridges, and aircraft cockpits to operator displays on medical equipment, manufacturing machinery, and construction equipment; the design of control rooms for power plants, chemical processing facilities, airport control towers and so on. In widely varied technologies and situations, engineers must apply principles that make the correct thing easy to do and faulty decisions difficult or impossible to execute.

Human factors analysis takes into account many physical and psychological capabilities of operators: vision, hearing, touch, and so on. It must ensure that control rooms are designed so that displays are properly lighted, large enough to read, and organized in such a way as to not be confusing. Designers must pay attention to possible interference from the volume and number of voice communications, to the frequency of alarms or other warning devices, and to the background noise, whether from machinery operating or people talking.

Where the physical forms of controls are similar, ways of hurriedly distinguishing among them in stressful situations often can prevent an operator from pressing the wrong button or throwing the wrong switch. Contrasting colors make similar controls more readily distinguishable. Likewise, distinctly different shapes may reduce errors: if two levers must be next to each other, making the handles in the shapes of a sphere and a cube, for example, will allow an operator to feel the difference between them, even if she has to look at a display in another direction.

Standardization in location is also important, particularly where operators have to switch between different locations or brands of equipment. Recall your last experience renting a car. Hopefully you checked the location of the lights, horn, emergency brake, windshield wipers, and defoggers—perhaps trying them out—before you set out on your trip. Although the details may differ, the locations are relatively standardized, and for good reason. If you need them in an emergency, and your hands are in the wrong place, an accident might well result. The location of the light switch was standardized because of just such experiences.

More than a half century ago, following the completion of the Pennsylvania Turnpike with its many tunnels, a number of fatal accidents occurred that at first were inexplicable. However, accident investigations pointed to the fact that the autos entered the tunnels without their lights on, and that they were rented cars! Further analysis revealed that the car renters expected the light switches to be located in different places on the dashboard, namely where the switches were in the cars they owned. The investigation of the tunnel accidents led to an agreement between auto manufacturers to locate the switches in the same place in all of their products, greatly reducing the hazard. Of course, this particular problem has been circumvented over recent decades, either by having lights on under all conditions or by having sensors that cause them to come on as soon as visibility is reduced. But the locations of emergency brakes, windshield wipers, defrosters, heat and air conditioning, and so on are also important.

Obeying population stereotypes, i.e., making controls intuitive and obeying ingrained habits rather than making operators learn new ones, also reduces the chance of error. Thus convention dictates that red and green imply stop and go; "on" is "up" on a light switch, and numbers increase clockwise on a dial. But conventions vary, even between professions; this must also be taken into account in interface design. For example, to turn electrical equipment on, turn the dial clockwise. But for someone used to dealing with hydraulic equipment, you turn the knob counterclockwise to increase the flow of a fluid. To turn the volume up on a radio, twist the dial clockwise, but to turn on the water for a garden hose, twist the faucet counterclockwise. In panicky situations, such differences may lead to inappropriate reactions, with operators reverting to deeply ingrained habits. The challenge for human factors engineers is to design controls in ways that eliminate the possibilities of such confusion.

With internationally employed technology—in marine and air transport, for example—the problem of human error caused by clashes between cultural norms is compounded. It goes beyond communications between captains or pilots who speak different languages, or even the same language with different dialects or accents. On a US light switch, "up" is "on", but in Europe, "down" is "on", as the author discovered one night while trying to attend to a crying child in a European hotel.

Anyone from the United States who has had to drive a rental car in the United Kingdom, or for that matter in in a number of other countries, must go through a rapid learning process to automatically put their hands in the correct place to shift gears, turn on windshield wipers and so on. Even stepping off a curb to cross a street requires first glancing in the opposite direction from the reflexive action that we have had ingrained since childhood.

Making technological interfaces clear and readily understandable frequently comes into conflict with other priorities. New innovative designs must be weighed against the errors that operators may make if they are not thoroughly familiarized with the new way of doing things. New and even clearer controls may improve safety, but only if the operators are thoroughly accustomed with their use. Aesthetics in particular can be the enemy of technological interfaces designed to reduce the risks of human error. Changing fashions and stylistic innovations, whether they be in light switches, automobile dashboards, or power plant controls, may lead to designs that are more pleasing to the eye at the expense of simplicity and familiarity that may be critical.

Clarity may be sacrificed for aesthetic reasons. A classic example of this appeared in the design of the Dresden nuclear plant outside Chicago, built in the late 1960s. Two reactors shared the same control room, with the control panels located next to each other. Apparently for aesthetic reasons, the designers configured the controls for the pair of reactors to be mirror images of each other, which made for a very symmetrical, nice-looking control room. However, on occasion the operators for one unit may have needed to switch to the other. But then what would happen? The controls that they would expect to be on the right would now be on the left, and vice-versa. This would impede their reactions if an emergency occurred that required fast thinking and decisive action by the operators.

* * * * *

In contrast to the continual decision-making required in many areas where technology is applied, some people in positions of responsibility may spend long periods of time simply monitoring equipment to assure that things are going well, punctuated by more intense periods of activity. With the highly automated controls in modern airliners, for example, pilots have relatively little to do while in level flight, with their activity concentrated during takeoff and landing. Likewise, operators of chemical refineries, power plants and other industrial facilities that operate 24 hours a day, seven days a week face long periods of routine monitoring of equipment—and even boredom—unless startup, shutdown, or changes in power throughput or other operating conditions are underway.

More important, emergencies may arise on rare occasions, whether caused by an aircraft's failed engine or a ruptured cooling pipe in a chemical processing plant. In such circumstances, those in charge must spring into action, making the decisions that will bring the situation under control, mitigating damage, and preventing loss of life. Equipment designed to assist operators under emergency situations is essential in helping them to diagnose the problem and take appropriate action. If it is not, the results can be disastrous. Nowhere is this better illustrated than in the accident that took place on March 28, 1979 at the Three Mile Island nuclear power plant.

In a nuclear reactor such as the one at Three Mile Island, water at high pressure and temperature carries heat produced by nuclear fission away from the steel vessel in which the reactor core is encased; the heat is then converted to electricity. A mechanical equipment failure, compounded by maintenance errors, interrupted this flow and initiated the Three Mile Island accident. The result was that heat could no longer be removed from the reactor vessel through the normal pipeways.

The disruption automatically drove control rods into the reactor core, shutting down the nuclear chain reaction that causes uranium to undergo fission. Even though the chain reaction had been terminated, a few percent of the reactor's power continued to be generated from the heat produced by the fission products' radioactive decay. To remove the decay heat, the reactor core must be kept covered with water. If it isn't, the uranium fuel will begin melting within four or five hours.

The reactor was equipped with an emergency cooling system to circulate enough water though the vessel to remove the decay heat. If water is pumped into the vessel more rapidly than it is removed, however, it could cause the pressure to rise high enough to rupture piping, or worse, the reactor vessel itself. To deal with the possibility of overpressure, the emergency system included a relief valve which was to open if the pressure became too high, and to stay open only long enough to bring the pressure down to an acceptable level.

The emergency system hardware had sufficient capability to deal with the events at Three Mile Island. But confusing controls and display panels, coupled with the operators' lack of training in dealing with emergencies, exacerbated the situation and turned what would have been a minor incident into the worst accident experienced by a power reactor operating in the United States.

The operators' confusion resulted from two control room indicators, one for the relief valve and the other for measuring the water level in the reactor vessel. The relief valve requires power to stay open. The problem was that the indicator in the control room was lighted only when the power to the relief valve was on. If the valve was operating properly, this was fine: light on = valve open, light off = valve closed. Shortly after the accident began, the valve opened to relieve pressure, but it became stuck in the open position.

As the pressure in the reactor vessel dropped, the indicator light went off since power to the valve was no longer needed; therefore, the operators thought the valve had closed. But it hadn't. At the same time, the reactor vessel gage indicated that the water level was rising rapidly. But this too was misleading, for in fact the water in the reactor vessel was boiling. A similar phenomenon can be seen when a pot boils over, the water level seems to rise even though no water has been added.

The two indicators seemed consistent, but extremely worrisome. They implied that since the light was off, the relief valve must be stuck in the closed position, bottling up the reactor vessel. At the same time, the water level gage indicated that the vessel was filling rapidly and would soon overfill, causing over-pressurization that might destroy piping or the vessel and render it impossible to cool the reactor core.

In the confusion caused by the faulty instrument readings, the operators inappropriately used a manual override to turn off the emergency cooling system. In fact, only the water supplied by that emergency system was compensating for the water escaping through the stuck-open relief valve. The operators were not able to figure out what was really going on for several hours, until the next shift of workers arrived and made an independent assessment of the situation. The delay was crucial, for without the emergency cooling water, the reactor's core had become uncovered, and much of it eventually melted.

Fortunately, the containment building that enclosed the reactor system functioned as designed. It confined the substantial amounts of radioactive material that were released with the steam and water during the course of the accident. Only minuscule amounts of radioactivity escaped to the atmosphere, and no one died as a result of the accident. The Three Mile Island accident accentuates the importance of control room design that ensures unambiguous indications of the true state of the plant to the greatest extent possible. The confusion in the control room also pointed to the need for comprehensive operator training, so they can diagnose problems, determine if some of their instruments are malfunctioning, and take correct and decisive action.

A great deal of effort has been made since the accident to improve control room design to make misdiagnosis less likely. Job preparation and periodic review courses now make extensive use of simulators that mock up control rooms and present the operators with multiple accident scenarios, including those with instrument failures. Prominent among the lessons learned from the accident was that equipment design and operator training must go hand-in-hand to ensure that when crises occur, human intervention will lessen and not exacerbate the adverse consequences.

* * * * *

Poorly designed interfaces between humans and their technology have combined with inadequate training to produce much more serious accidents than that experienced at Three Mile Island. The fate of Air France Flight 447 offers a sobering example. On June 1, 2009, the Airbus A330-200 took off from Rio de Janeiro, Brazil, on its way to Paris carrying 228 passengers and crew. But then it disappeared over the Atlantic, without even a distress call from the pilots.

Authorities launched an intensive search along the aircraft's flight path; on the following day, the first debris was spotted floating on the ocean. However, it was nearly two years before the searchers recovered the black box flight recorder from the three-mile deep waters of the Atlantic. Only then was sufficient evidence available to piece together what likely happened.

The authorities suspected that ice crystals forming in the airspeed sensors as the aircraft passed through an extended region of towering thunderstorm clouds caused the accident. The ice buildup caused the sensors—called pitot tubes—to malfunction, producing faulty readings that indicated that the Airbus had lost much of its speed. The aircraft had been flying on autopilot, monitored by the copilots while the captain rested. The faulty and inconsistent airspeed signals caused the autopilot system to disconnect, turning the plane over to manual control of the copilots.

The disconnect left the copilots attempting to control the plane while concentrating on the untrustworthy airspeed readings, thinking that they were accurate. Their situation was further complicated by faulty instructions and a profusion of alarms generated by an automated navigational aid called the flight director. Using data from the faulty sensors, the flight director computer repeatedly instructed the pilots to lift the plane's nose, which they did. But this put the aircraft into a sustained stall.

As shown in Fig. 5.1, a stall occurs when an aircraft's nose is pointed too far upwards. The wings then lose their lift because of a chaotic air flow pattern that develops around them. This is what investigators determined to have happened. Instead of pointing the nose downward, which is the standard maneuver to come out of a stall before it becomes too severe, the pilots continued to push the nose upward until they passed the point of no return. The aircraft's wings lost all lift, and for three horrific minutes, the aircraft plunged from an altitude of nearly 40,000 feet down into the Atlantic, taking all 228 passengers and crew to their deaths.

Who or what was at fault? Certainly equipment failure, faulty human interface design, and inadequate training of pilots to deal with

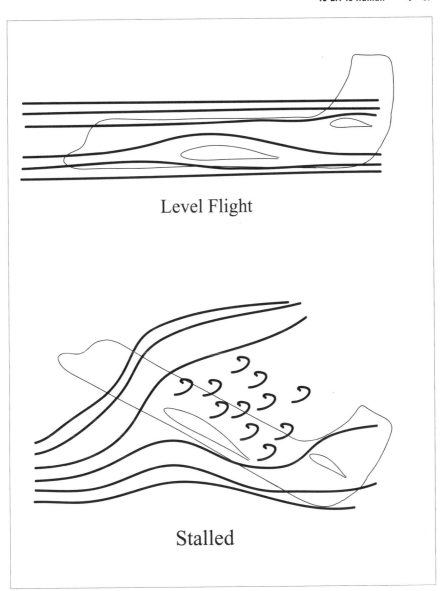

Level Flight

Stalled

Figure 5.1: Aircraft Stall Conditions

emergencies combined to cause the disaster. There had been earlier but rare instances of pitot tube icing, at least two of them causing fatal crashes. Thus it would seem that the flight director system, designed to assist the pilots when they were flying the aircraft manually, should have been programmed to take the possibility of airspeed sensor failures into account, and not display inappropriate or inconsistent instructions during an emergency.

The pilots had no direct indication that faulty speed sensors were the root of the problem. In their increasingly panicky state, and listening to a flurry of conflicting alarms and instructions, their training didn't provide them with the capability to work their way quickly through the conflicting information displayed on their control panels, diagnose the underlying problem, and take appropriate action.

The investigators' final report on the tragedy recommended improvements in the flight director software to eliminate confusion, so that what is supposed to be a navigational aid will not issue misleading or contradictory instructions in emergency situations. They also recommended enhanced pilot training on cockpit simulators to increase pilots' ability to diagnose and deal with emergencies stemming from failure of sensors or other equipment. Importantly, the investigators also pointed out that, with the increased automation of flight controls, pilots' skill in flying aircraft manually under adverse conditions is diminished, and with it often their ability to quickly deal with equipment failures, high altitude stalls, and other emergencies. Specifically, the report stated, "We are seeing a situation where we have pilots that can't understand what the airplane is doing unless a computer interprets it for them. [...] It's a new training challenge that the whole industry has to face."

* * * * *

The design of automobiles is evolving rapidly, analogous to the changes that took place in commercial aircraft in the last decades of the twentieth century. Fly-by-wire replaced the hydraulic and mechanical links by which pilots once controlled aircraft with computers and sensors that changed the feel of the aircraft to the pilot flying it. Now a similar process of change has taken hold with automakers. Antilock brakes and stability control systems are already in use. The driver's relationship with the steering wheel and the gas and brake pedals is also changing. Electronics has taken over the connection between the gas pedal and fuel injection; this mechanical linkage has been replaced with electronic throttle control. And in the works are brake-by-wire and steer-by-wire.

When the transition is complete, autos will have become drive-by-wire systems. Aircraft designers implemented fly-by-wire in ways so that they kept as much as possible the feel of flying the aircraft that pilots had with the traditional mechanical links. They thought it was a good idea to minimize pilot errors—some of them potentially fatal—if pilots had to suddenly adjust to a control system with an interface that was radically different from the one with which they had gained experience. For safety's sake, we can only hope that automotive designers will pay close attention to the motorists' need for continuity.

In commercial aircraft, changes in the relationship between pilots and their machines have progressed far beyond replacing mechanical with electronic linkages. Autopilot controls and other advanced systems can take off, fly and land the aircraft without human intervention. And while such capabilities help relieve pilot fatigue and have other safety advantages as well, they also present challenges. As we saw in the tragedy of Air France Flight 447, pilots must be able to diagnose faulty instrumentation and take corrective action if the automated flight control system encounters a situation that it has not been programmed to handle.

Such situations aren't unique. This became apparent in the 2013 crash landing of a Boeing 777 in San Francisco. The airport's instrument landing system was shut down for maintenance. Yet even though the weather was clear, the pilot, unaccustomed to making manual landings in that aircraft, came in too low and slow. Consequently, he clipped a seawall with the aircraft's landing gear and tail. The wide-body jet slammed to the ground, skidded down the runway, and then caught fire. Fortunately, the aircraft's emergency evacuation system worked well. Had it not, the death toll among the 309 aboard would have been far greater than the two lives that were lost.

The safety issues encountered with automating aircraft controls raises the question of what the effect on the death tolls from motor vehicle accidents would be if predictions come true that, in the near future, automobiles will be able to drive themselves while the passengers read, look out the window or talk on the phone. Proponents of such systems point to many safety advantages, since the automated system compensates for changing road conditions, traffic patterns and more.

But what unanticipated emergencies might arise—sensor failures or bizarre unanticipated traffic configurations, for example—that would cause the system to shut down, much as the autopilot did on Flight 447? Would the driver be able to take control and deal with the situation manually? It seems unlikely that all 50 state Departments of Motor Vehicles would require simulator training to deal with such situations before buying such a vehicle. So we can wonder what licensing requirements will be like in an age of self-driving automobiles.

Hopefully, self-driving failures will be rare and result in few, if any, lost lives. If risks are outweighed by the positive effect of reducing the number of motorists trying to drive manually while drowsy or under the influence of alcohol, self-driven cars may reduce the death tolls on our highways.

I was thinking of these tradeoffs while driving a car I recently bought. While the vehicle was still far from futuristic automated driving, it did incorporate many advanced technological interfaces. Some were first introduced to assist airline pilots and ship captains and have been adapted to motor vehicles. The dashboard had little resemblance to those in autos of less than a decade ago. A sizable computer screen and an array of five dials and more than thirty buttons and keys replaced the relatively traditional arrangements in older cars. There was also GPS with a continuous map display, keyless ignition, satellite radio, more elaborate cruise control, wireless cell phone connections, and much more that I would have not dreamed of using only a few years ago.

I hadn't yet figured out how to operate all of the electronic gadgetry, and I had to force myself not to fiddle with it while driving. There is a danger of being distracted by too many added features, but hopefully the auto's added safety features will outweigh those potential dangers. The backup camera decreases the likelihood of hitting a small child in the blind spot below the rear window, and also makes parallel parking a great deal easier. I particularly appreciate the surveillance system that activates lights adjacent to side-view mirrors if another vehicle has entered the auto's corresponding blind spot. With them, I can more confidently change lanes or pass another car without twisting my neck to check the area not visible from the side view mirror.

But then I began comparing the mirror to the electronic indicator light. The mirror has the disadvantage of not offering a full view, forcing me to remember to look over my shoulder before changing lanes. But the mirror has the advantage of signaling me if it fails. I know immediately if the mirror fogs over or is out of position, showing sky or countryside instead of the road behind me. In contrast, if the sensor or warning light fails, there may be no indication that the system is no longer working. I must check frequently as cars over-take me (or I overtake them) that the light goes on at the appropriate time, indicating that the system is working. This may seem a simplistic comparison, but it demonstrates the issues that arise as we go to more virtual electronic interfaces to connect humans to their technology. It points to safety questions that must be addressed if advances in convenience and comfort shouldn't create new hazards that increase technological risk.

CHAPTER 6

Cultures of Safety—and Lack Thereof

Densely populated shanty towns surrounded the large industrial plant in Bhopal, India. Owned by the Union Carbide Corporation, it produced pesticides vital to increasing the country's agricultural production. The plant, however, was plagued by a variety of shortcomings in its design, operations, and safety precautions—many of them exacerbated by conditions in the developing world, cross-cultural barriers, and the parent company's hands-off attitude toward its foreign operations. The safety equipment and operating procedures employed in Union Carbide's US sister plant located in Institute, West Virginia far exceeded those required at Bhopal.

The hazardous condition of the Bhopal plant and its operations were known to the local authorities, but—as in countless other situations in developing counties—the desperate need for economic development outweighed concern for safety; the expense of safer operations might cause one of the city's largest employers to move its operations elsewhere.

The situation in the plant was indeed abominable on the night of December 3, 1984. There were too few operators, and most of those lacked proper qualifications. Large, poorly maintained tanks of methyl isocyanine (MIC) were filled beyond recommended levels. Mechanical equipment—pumps, pipes, and valves—were in disrepair and frequently became stuck or clogged, then malfunctioning. Safety systems were not in service, some from lack of maintenance but others simply because managers had turned them off to save money.

A disastrous situation developed in the early hours of the morning when the combination of a leak and a faulty valve allowed water to mix with 40 tons of MIC in a storage tank. Immediately, a violent chemical reaction built upon itself, generating huge amounts of heat and pressure. Safety devices for neutralizing the toxic mixture had been turned off weeks earlier. Thus pressure built relentlessly toward the tank's bursting point, causing a safety valve to spring open and release a lethal plume to rumble into the atmosphere and spread over the city of 900,000.

Over the course of less than an hour, a cloud of highly toxic gas spewed forth from the plant, engulfing the surrounding area's residents as they slept. There was no warning, for the plant's emergency sirens had been turned off. Thousands died in their beds, and streets near the plant became littered with human corpses and the carcasses of cows, dogs and birds. Nearly 4,000 residents died immediately. Hospitals were overwhelmed and further hindered because they had no idea what chemical was at fault. Estimates of the death toll in the following days ran as high as 10,000, with 50,000 more suffering permanent disabilities. The Bhopal cataclysm was the worst industrial accident to occur, ever, anywhere in the world.

* * * * *

The disaster at Bhopal didn't occur in an industrialized democracy, nor did the worst disaster involving nuclear energy. Rather, that accident occurred in the Soviet-dominated Ukraine toward the end of the Cold War. It took place at Unit 4 of the Chernobyl Nuclear Power

Plant on April 26, 1986. The precise causes of the accident remain somewhat in dispute, both because many valuable instrument readings were destroyed with the accident and because of the reluctance of the Soviet government to release what records it had. But clearly both the reactor's design and the manner in which it was operated led to the disaster.

The Soviet-designed reactor was not encased in the protective containment vessel, as required in the West, as a final barrier to separate the radioactivity in the reactor's core from the outside environment. Equally important, the Soviet-designed reactor had shortcomings in its peculiar control system and inherent instabilities at low operating powers. These would not be allowed under the regulations to which reactors must conform in the developed world outside the USSR. Finally, because of insufficient training, the operators didn't understand the design's inherent dangers well enough to take adequate precautions in performing their duties. Instead, they disconnected the automatic shutdown devices from the reactor before they continued with the ill-fated low-power tests, which had not been authorized or reviewed by more senior managers or engineers.

Everything went wrong in the operators' attempts to manipulate the unstable reactor. Making on-the-spot changes in the test procedures, they set off an explosive power surge of unprecedented proportions that destroyed the reactor core and blew the top off of the building housing it. The graphite that constituted a large part of the reactor core ignited, and the fire that ensued sent a highly radioactive plume into the atmosphere that spread contamination over large areas of the western Soviet Union and Europe.

The death toll in the weeks following the accident was much smaller than for the Bhopal disaster. Acute radiation exposure took the lives of 29 plant workers and sickened many more. To those living in the expansive contaminated areas outside the plant boundaries, the longer-term risk of cancer stemming from lower levels of radiation exposure constituted a much more pervasive health problem. With one glaring exception, however, estimates of cancer incurred from the accident's radiation have amounted to less than one percent

Figure 6.1: The Chernobyl Nuclear Power Plant, Unit 4, Following the Accident, *Associated Press*

of the normally occurring rate, an increase too small to detect. That exception was thyroid cancer.

Children and adolescents are particularly vulnerable to it if they ingest or inhale radioactive iodine, which then collects in their thyroid glands. In the 25 years following the accident, 6,000 thyroid cancers were reported, a number that may grow in coming years. Fortunately, with proper treatment, the prognosis for thyroid cancer is very good; to date, only 15 fatalities have resulted, a small number compared to the thousands who perished in the aftermath of the chemical spill at Bhopal.

The deficiencies in the safety culture that existed under the Soviet dictatorship has much to say about the severity of the accident, caused by lax reactor design combined with incompetent plant management. Those deficiencies tell even more about the thyroid cancer epidemic that followed. The Soviet government tried to conceal the accident from the world and its severity from those who lived near the reactor. For hours following the accident, the people of the nearby town of Prypiat were oblivious to the dangers that they faced. Only after elevated radiation levels in the atmosphere set off alarms in Sweden, many hundreds of miles away, did the accident become known to the world; widespread reporting in the Western press forced the Soviet Union to admit to its own people that the accident had occurred.

In the following months, over 100,000 people were evacuated from the communities nearest the reactor; eventually, even more were relocated. But the pace was slow. Even more important: if an effective system of emergency planning had been in place, using basic public health measures could have significantly reduced the 6,000 thyroid cancers.

The danger of ingesting radioactive iodine lasts only a few months; after that, virtually all of the radioactive iodine has decayed away. But during those months, the iodine settled on agricultural land, was eaten by cows and concentrated in their milk. If milk from the contaminated areas had been quarantined immediately following the accident, the children's ingestion of radioactivity would have been much less, and exposure for the thyroid gland greatly reduced. With adequate preparation, other measures would also have helped. If iodine tablets had been taken immediately following the accident, they would have blocked the thyroid's uptake of radioactive iodine from contaminated milk—even it had been drunk—and from other contamination as well.

* * * * *

Not all man-made disasters result from high-tech ventures such as those at Bhopal and Chernobyl going awry. Train wrecks, ferry boat collisions, fires, and building collapses all too frequently result in large loss of life. Two of the most deadly illustrate the point.

On the night of December 20, 1987, the *Dona Paz*, a Philippine inter-island ferry, overloaded with sleeping passengers, was returning to Manila for the Christmas holidays. As the passengers slept, the ferry boat collided with a tanker carrying tons of gasoline, kerosene, and diesel fuel. Both ships caught fire immediately, trapping most of the passengers; within two hours, the *Dona Paz* sank, as did the tanker two hours after that.

The disaster was worsened by the state of the two ships. The tanker was operating without a license, a lookout, or a qualified captain. Reports indicate that the lifejackets on the *Dona Paz* were locked in lockers, and the ferry had no radio! Without radio contact, eight hours elapsed before authorities got word of the disaster, greatly delaying rescue attempts. The legal capacity of the *Dona Paz* was 1,500 passengers, but the ship was greatly overloaded. Later investigations put the collision's death toll at 4,386; only 24 from the ferry survived, along with two of the tanker's 13 crew members. It was the most deadly disaster in peacetime maritime history.

More recently, in April 2012, disaster struck in Bangladesh, where thousands of workers were crowded in building housing several garment factories. The collapse of the Rana Plaza Building in Savar, a suburb of Dhaka, was not without warning. As typically happens with the lack of enforceable regulation often found in poor countries, the building's construction was shoddy. In addition, the upper three floors had been added illegally after the initial construction, without consideration of the stress that the added weight would place on the structure. More immediately, on the day before the disaster, an engineer discovered an ominous crack in the building and ordered evacuation, stating that the structure should not be reoccupied until it had undergone a thorough investigation.

Time pressure to produce low-cost garments for Western apparel companies overwhelmed safety considerations: on the following morning, a bank and some shops on the lower floor of the building remained closed. But owners of the five garment factories on the upper floors ordered their employees back to work, implying that the building was safe. Packed with workers, the building collapsed with deafening noise and a choking cloud of dust. Weeks of digging through the rubble to rescue the living and find the dead followed. The toll rose to more than 1,100 deaths and 2,500 injuries, surpassing previous fires and collapses as the worst accident in the history of the garment industries.

The foregoing disasters caused death and destruction on such a scale that they attracted worldwide media coverage, bringing attention to the appalling lack of safety standards in what are known as second- and third-world countries. But disasters of such proportions account for only a small fraction of the appalling accidental death tolls in developing countries. Many of the numerous accidents that take only one or two lives at a time relate to the introduction of technology into cultures that are too poverty stricken to apply it safely.

Present-day technology surely makes inroads to lifting the standard of living, but the conditions under which it is used and the state in which it is maintained are dreadful. Trucks may have replaced many oxen, camels, or donkeys, but they are overloaded with occupants hanging from the sideboards and in danger of falling off. Vehicles ply rugged tracks and poorly maintained roads with frequent breakdowns caused by frayed tires, overheated engines, and so on. Invariably, the trucks are in terrible disrepair—with spare parts likely to be as unavailable as are technicians capable of installing them. No wonder that, per mile traveled, the death rate from motor vehicle accidents is much higher than in the industrialized world.

Cooking on kerosene stoves may be a welcome improvement over open fires, but the stoves are prone to tip, spill and engulf crowded shanties in flames. Likewise, where governments have been able to string electric lines to provide lighting and power to pump potable water to the residents of city slums or rural villages, those lines are likely to quickly become overloaded—overridden with jerry-rigged connections that are both dangerous and illegal. But in their desperation for light and heat, the inhabitants are likely to resist—sometimes violently—the efforts of authorities to limit distribution to those who can be safely served.

In such situations, the people are so desperate for the improvements that technology brings that they adopt it readily. Unfortunately, the very poverty they seek to escape does not provide for the training, maintenance, and infrastructure necessary to counter the safety hazards inherent with the use of technology—seat belts are a luxury in Somali, and smoke detectors nonexistent in Darfur. Even in areas where safety regulations are in place, corruption frequently undermines their enforcement.

* * * * *

As witnessed by the disasters in India, the Philippines, and Bangladesh, grinding poverty, inadequate infrastructure, and weak governance make technology's use more deadly in the developing world than in the industrialized democracies. Likewise, even where industrialization is more advanced, if democratic institutions do not ensure responsive government, deficient safety cultures may cause catastrophes, such as the one that occurred in Chernobyl.

The cultures of industrialized democracies tend to be more safety-conscious than those of poorer counties because higher living standards have negated the need for adopting technology under conditions that are inherently dangerous. Still, disasters and everyday accidents alike also take place even in prosperous societies, but they occur less frequently and take fewer lives. When disasters do occur, careful examination invariably reveals a safety culture that has fallen short. Read any blue-ribbon report written by a high-level panel commissioned to investigate a major disaster. An inevitable conclusion of such studies focuses on the need to improve the cultures of safety within the responsible institutions and often of society as a whole.

Safety depends on more than the design and manufacture of the technological equipment upon which it is based. It is also the result of the skill, commitment, and understanding of those who operate and maintain that equipment. And this in turn depends even more on the organizational culture in which they work.

Just as design engineers must make tradeoffs between performance, cost and safety, organizations that operate and maintain highly technological systems must continually make similar tradeoffs.

And when safety succumbs to pressures to reduce costs or increase performance, disaster may eventually follow.

Even within highly developed economies, cultures of safety vary substantially between institutions, whether they be corporations, government bodies, or community associations. Government bodies may not have the resources or the will to safely maintain the dams, highways, or other infrastructures for which they are responsible. Corporations are prone to give top priority to holding down costs and meeting deadlines, which often loom larger in the minds of managers than the seemingly remote possibility that something may go terribly wrong.

Whether the institutions are public or private, deficient cultures of safety may develop all too easily: laxity develops in attending to safety issues, in uncovering design flaws or in ensuring that maintenance and repairs are properly performed. Employees do not report problems that they discover if they fear that higher management will not be receptive, or worse, if it figuratively "shoots the messenger" by diminishing the employees' career prospects or even firing them and blackballing them from obtaining other positions. What follows are two disastrous scenarios relating to deficient cultures of safety in the United States. In the first, the failure to develop an adequate culture of safety falls primarily on the complacency and bureaucratic bungling of governmental institutions; in the second, corporate greed and mismanagement are the primary villains.

* * * * *

When Hurricane Katrina slammed into New Orleans on Monday, August 25, 2005, it caused the levee system to fail catastrophically; this resulted in the flooding of most of the city. Many of its residents were stranded on rooftops, desperately awaiting help; more than 1,500 lost their lives. By the time the storm had dissipated, 400,000 had fled and 100,000 were left homeless. The hurricane resulted in the worst humanitarian disaster suffered on US soil in nearly a century, and its causes were many, attributable as much to the lack of a culture of safety as to the intensity of the storm.

But it seems that it shouldn't have been that way. The lesser amount of devastation caused by Hurricane Betsy in 1956 had caused Congress to pass the Flood Control Act of 1965, which mandated the Army Corps of Engineers to protect New Orleans from "the most severe meteorological conditions considered reasonably characteristic for that region." It was thought that protection stemming from that legislation would require 13 years to complete; in fact, under-budgeting, bureaucratic delays, and piecemeal construction caused them to be stretched over 40 years, with estimated completion not for another decade. Worse, the level of protection fell far short of what was stipulated by the 1965 legislation.

The levee system design was based on a 1959 estimate of the magnitude of a once-in-100 year hurricane. The estimate was already known to be obsolete as the improvement project got underway. Ongoing studies indicated that the wind speed value used by the Corps of Engineers was 20 percent too low, leading to a 40 percent underestimate of the storm surge height caused by the hurricane. Even though the New Orleans District of the Corps was aware of these deficiencies by the late 70s, it didn't revise the strength of the 100-year hurricane that the system was designed to withstand. The Corps continued to accept the inadequacy of the levee system.

Without supporting evidence, the Corps, in its complacency, repeatedly claimed that the system under construction would survive even a once-in-200 year or once-in-300 year hurricane. Also among its failures, the Corps didn't account properly for weaknesses in the soils beneath its floodwalls, nor did it compensate for the well-understood ground subsidence that would take place after the structures had been built. These erroneous assumptions resulted in floodwalls and levees that were more than two feet too low and prone to failure.

As the project fell further and further behind schedule, confusion grew: maintenance and upgrade responsibility could not be passed to local authorities, as the original plan had stipulated, nor were local governments consulted on design and construction decisions and revisions. Without local collaboration, the design was inadequate, the construction was faulty, and the system was poorly maintained. When Katrina struck, it exceeded what the Hurricane Protection

System was designed to withstand, and the construction and maintenance deficiencies caused the system to fail at even lower levels.

The storm damaged nearly 200 miles of the levee system, with breaches opening at more than 50 locations. Four floodwalls failed even before the water reached levels that they were designed to withstand. The breaching and erosion of the levees increased the flooding by 300 percent of what it would have been if the walls had held, and only the water overflowing their tops had entered the city.

It was not only the gravely deficient levee system that caused the loss of life, livelihood, and property. The storm proved the organizations at the city, state, and federal levels that were empowered to protect the public to be dysfunctional. But this was not for lack of warning; the expected storm severity was well understood by the responsible government agencies. A few years earlier, the Federal Emergency Management Administration (FEMA) had conducted a multi-agency planning exercise, dubbed Pam, for dealing with a severe hypothetical hurricane.

Poor and delayed political decisions, as well as lack of coordination between government bodies, led to an absence of an executable plan to move the population away from areas that would flood if the levee system failed. Once the storm hit, the lack of planning resulted in unconscionable delays in rescuing those who were stranded and in getting aid to those who had fled. Clearly, the emergency drills needed to identify and remedy gaps in knowledge, poor communications, and fuzzy lines of authority had not taken place. The defective plans that did exist didn't result in effective evacuation before the storm or in timely recue and recovery efforts afterward.

Although New Orleans's emergency plan had contingencies for quickly evacuating threatened areas and rescuing large numbers of stranded people, little of the plan was adequately executed when the storm came. Local authorities inexplicably delayed evacuation orders for more than 24 hours after evacuation was strongly advised by federal meteorologists at the National Hurricane Center. Nor were New Orleans's residents prepared to comply, and many resisted leaving until it was too late.

Confusion was rife in coordinating the efforts of local, state, and federal authorities and of private relief agencies. Excessive regulations, bureaucratic fumbling, late and misdirected requests by the state, and a lack of proactivity on FEMA's part crippled both the emergency response and the reconstruction efforts to follow. Among those in the direst circumstances were the many who were directed to the Superdome and the Convention Center. They were stranded there for an unconscionable length of time, with no provisions made for security, food, or water. With much of the population homeless, thousands of FEMA trailers remained unused in the weeks following the storm, and unresolved contracting questions added to the dysfunction, delaying reconstruction efforts for much longer than should have been the case.

The lack of a culture of safety that led to the tragedy in New Orleans was not due to a single cause, but to complacency at many levels of both governmental and private agencies. The Corps of Engineers inability to update and push for the rapid construction of a levee system that could withstand what climatologists considered the best estimate of a once-in-100 year hurricane was critical. Worse, if a hurricane worse than the estimated 100-year storm hit the city and overtopped the levees, the levee design didn't prevent their total failure from catastrophic breaches.

There were many who shared responsibility for the tragedy: Congress for underfunding levee improvements and emergency planning, governmental officials at all levels who did not take seriously enough the need for rapid and reliable communications or the need to bring in aid in specific forms to specific places, and finally those who failed to plan the contracting process so that when disaster struck, delays in the delivery of aid would be prevented.

* * * * *

On March 23, 2005, a disaster struck BP Corporation's Texas City oil refinery, the third largest in the United States. It was the country's worst industrial accident in recent decades. The problem stemmed from workers filling a large splitting tower, employed for increasing gasoline octane ratings, and using a highly combustible liquid. With the warning alarm disabled, employees overfilled the container, and their rushed attempts to contain the situation made matters worse. A flammable liquid geyser spewed from the apparatus, and a spark from a nearby truck ignited it. The explosion and the ensuing fire killed 15 workers and injured 180 others. Nearly 50,000 people in the surrounding communities were ordered to remain indoors following the explosion for fear of toxic fumes; the blast wave damaged buildings as far as three-quarters of a mile away. The financial losses exceeded $1.5 billion.

Investigations following the disaster condemned BP's lack of a safety culture, asserting that organizational and safety deficiencies at all corporate levels were responsible for the disaster. Corporate management had ignored the warning signs of possible disasters that had been apparent for several years: increasingly dilapidated equipment, accidents fatal to one or two workers, and lax attention to safety by the refinery's employees.

Two serious incidents just months before the explosion further pointed to the grave deficiencies in BP's safety culture. They each resulted in millions of dollars of damage and orders for people in the surrounding community to remain indoors. At about that same time, a consulting firm documented BP's "broken alarms, thinning pipe, chunks of concrete falling, bolts dropping 60 feet, and staff being overcome with fumes." Still, corporate management made no efforts to prevent the eventual outcome.

Following the disaster, the US Chemical Safety and Hazard Investigation Board concluded that the corporate culture and structure had as much relevance to the accident as its immediate causes did. Simply putting blame on the operators and supervisors on site on the day of the accident would miss the point: the fault was due to the lack of an organizational safety culture that would have prevented the underlying causes from developing in the first place. The Board found that BP used inadequate methods for measuring the potential for catastrophic explosions or chemical release, focusing only on slips, trips, and falls—individual injuries measured by OSHA's workplace safety standards. BP neglected the steps that would lessen the growing risk of a catastrophic event.

The situation had steadily grown worse as BP's operational policies had dramatically altered the production-cost-reliability tradeoffs: corporate management reached for maximum production while at the same time aggressively cutting costs. BP targeted operating budget cuts of 25 percent in 1999 and another 25 percent in 2005. Meanwhile, the wear and disrepair of the process equipment and infrastructure at the Texas City refinery became worse. Higher-level management downsized staff and reduced operator training. And without repair and replacement, the equipment failure rates increased, leading eventually to disaster.

Texas City was not the only place where BP's lack of a corporate culture of safety had dire consequences. The firm also held major operating responsibility for the 800-mile-long Alaska oil pipeline in the 1990s and thereafter. As the productivity of the oil fields diminished, and with it the amount of oil passing through the line, BP headquarters demanded that operating costs be drastically cut. Just as in the case of the Texas City refinery, the firm was pushing to preserve its capital for what it considered to be the more lucrative adventure: deep-water drilling in the Gulf of Mexico.

The results were predictable. The corporation reduced the number of maintenance and safety personnel, and their training as well. Outside contractors were increasingly used to save money, and the contractors often hired uncertified personnel. Inspection of the pipeline system was reduced through the periods of time when corrosion was causing the walls of the high-pressure pipes to thin, increasing the danger that they would burst, risking explosions and spills. Safety systems for warning of problems—such as for detecting the onset of gas leaks and fires—and those for emergency shutdown were allowed to fall into disrepair, or were simply shut off. Management ignored personnel who reported problems. If they persisted in reporting safety hazards, higher management threatened them with firing, and in a number of cases those threats were carried out.

One ploy by which BP evaded responsibility was to contract out inspection, maintenance, and repair to smaller companies, putting the onus on them. And the difficulties where magnified in those companies, because employees were more easily fired if they persisted in raising unwelcome safety concerns. Once fired, the whistle-blowers had difficulty finding other employment, for few jobs outside the pipeline were available for the specialized skill sets that they possessed. A collusion between BP and its contactors to blackball whistle-blowers all but assured that they would remain unemployed.

The failure of BP to properly maintain its pipeline first drew national attention in 2006, when two spills from corroded pipes resulted in the temporary cutback of oil transport from Alaska's North Slope to the continental United States. But other problems were soon to follow. In August 2007, the flammable liquid discharged when a jerry-rigged oil hose ruptured, causing a giant gas-compressing turbine to burst into flames. The fire and gas detectors were out of service at the time. Had the uncontrolled fire that ensued detonated the nearby high-pressure gas and oil pipelines, the result would have been much worse.

In 2008 and 2009, the risk of three more explosions came about from ruptured or clogged pipelines on Alaska's North Slope. In September 2008, there was another close call when a high-pressure gas line erupted, spewing combustible gas into the atmosphere and flinging a 28-foot long metal casing 900 feet. Fortunately, the cloud didn't ignite. Again, in January of the following year, large amounts of gas leaked from a pump station as the result of faulty maintenance operation.

In a separate incident, a stuck valve at a Prudhoe Bay compressor station caused large amounts of gas to vent to the atmosphere. Once again, safety equipment was not operational; more importantly, the flare that was supposed to burn off the gas in a controlled manner as it escaped was not operational. A disastrous blast was avoided only because no ignition source was present, and the cloud dispersed in the atmosphere. A month after, another rupture occurred—this time from an 18-inch pipe that spewed forth its flammable contents.

Fortunately for BP, Alaska operations survived the near-misses of catastrophic explosions that had the potential to cause more fatalities than the Texas City refinery disaster. BP continued to operate the Texas City refinery and the Alaska oil pipelines on the cheap, extracting profits at minimum cost, and narrowly avoiding disastrous accidents. Their priority was to raise capital for deep-water exploration in the Gulf of Mexico, which they looked to as their road to future profitability.

Once ingrained, corporate cultures do not change without concerted effort by top management; abundant evidence indicates that BP's lack of a safety culture carried over to the Gulf of Mexico. It was responsible in no small part for the disastrous explosion of the Deepwater Horizon oil rig off the Louisiana coast and for the widespread environmental devastation flowing from the record-setting oil spill that followed. Blue-ribbon investigations pointed to deficiencies in planning and maintenance that went well beyond the actions of those on the oil rig when it exploded. Deficiencies began in upper management and resulted in what many consider to be the worst man-made environmental disaster in over a century.

Figure 6.2: The Deepwater Horizon Oil Rig Sinking, *US Coast Guard*

* * * * *

Experience demonstrates that reliable equipment and systems that take into account human factors are not enough to prevent disastrous accidents. Prevention depends at least as much on the culture in which the technology is employed. Studies point out the difficulty in creating cultures of safety. Within organizations, they require sustained commitment by senior management to establish positive worker attitudes and participation in making safe practices pervasive. Throughout an organization, emphasis must be placed on hazard identification, assessment, prevention, and control. A factor that is just as important as training employees to deal with emergencies is encouraging them to report potential hazards to their supervisors with the assured knowledge that the problem will we addressed, and they will not be reprimanded for slowing production or embarrassing those in charge.

Throughout much of the world, cultures of safety are exceedingly difficult to establish or maintain. In underdeveloped countries, where governments are most often weak or dysfunctional, or where civil unrest is rife, the toil of the people for daily survival, the struggle for food and shelter, and debilitating effects of disease overwhelm the attention that could otherwise be given to reducing the death tolls that have accompanied the introduction of even the most rudimentary of technologies. Even though standards of living may be higher, and the government strong and in control in dictatorships such as the former Soviet Union, technology's risks are likely to be ignored. For the officials of autocratic governments, there is little political price to pay for ignoring the risks that the citizens face.

Even in prosperous industrialized democracies, safety is culturally dependent. It depends on the government bodies, corporations, and other institutions responsible for the technology's safe deployment. Maintaining cultures of safety presents ongoing challenges. Lapses are frequent and sometimes catastrophic, as the foregoing examples of Hurricane Katrina and the Deepwater Horizon oil spill exemplify. Commendable safety cultures do exist, but they receive little media attention; their accomplishments, namely a paucity of accidents, keep them out of the news. An instructive example comes from the US Navy.

The US Navy's Nuclear Propulsion Program was headed for decades by Admiral Hyman Rickover, who instilled a culture of safety that has lived on long past his retirement. The program's officers, many of them with engineering degrees, and enlisted personnel undergo extensive training in the operation and maintenance of nuclear reactors. They study what could go wrong and participate in emergency planning, frequent drills, and refresher training.

Higher-ranked offices also are required to be nuclear-trained, including the captains of all nuclear-propelled ships, ranging from submarines to aircraft carriers. In addition to reporting to their operational commanders, each captain regularly communicates with the Director of Nuclear Propulsion, detailing the state of the ship's nuclear plant, needed maintenance, the status of crew training, possible manpower deficiencies, and other relevant matters. The Director of Nuclear Propulsion reports directly to the Secretary of the Navy, assuring that the safety of the nuclear systems has the attention of those at the highest organizational level.

There are other examples, as demonstrated by the "miracle landing" on the Hudson of Flight 1549. The pilots and crews of commercial airlines train and drill in preparation for emergencies, with the pilots spending hours in cockpit simulators to prepare them to deal with engine and landing gear failures, on-board fires, and other contingencies. At the local level, many fire and police departments, emergency room staffs and others train to prepare for dealing with rare but potentially devastating events.

Maintaining national cultures of safety extends beyond what is established in corporate and government bodies. It requires the participation of the public as well, both in their individual attitudes and actions, but also in how they support government bodies in regulating unsafe practices and behavior. Cultural attitudes toward the prevention of accidents vary substantially, even among prosperous democracies.

In some countries, automobiles must undergo periodic inspections of the brakes, steering, lights, tires, and other safety-relevant systems.

If the cars fail the tests, they must be taken off the roads until repairs are made. Consequently, many older cars are replaced much sooner than in places where law allows the owner to decide whether his vehicle is safe enough to drive. Likewise, some states require building owners to upgrade wiring, elevators, and other equipment to meet stricter safety standards, and to complete the work within a specified length of time. Meanwhile, others apply tighter standards only to new construction, but allow grandfathering; thus, long periods of time elapse before the tightened regulations have the desired impact in reducing hazards.

CHAPTER 7

Risks, Response, and Regulation

IN THE EARLY 1800s, engineers introduced compact engine designs that used steam well above the pressure of James Watt's historic invention. Steamboats powered by these high-pressure engines revolutionized transportation and brought about rapid economic development to America's western frontier. Overpowering the strong currents of the Mississippi and its tributaries, they facilitated the settlement of the fertile Middle Western states as well as the vast territory that the United States had recently acquired through the Louisiana Purchase.

A grave new risk, however, accompanied this revolutionary technology. As Watt had feared, high-pressure steam engines led to lethal boiler explosions. Within little more than a year of the engine's introduction, seven steamboat explosions resulted in 50 deaths. And things got worse: over the next 30 years, more than 200 explosions caused more than 2,500 deaths and almost as many injuries.

Then came a particularly grievous accident that garnered public attention and brought political pressure for change.

The seminal event took place in 1830. A boiler on the Mississippi steamboat *Helen MacGregor* blew up as the boat was leaving a dock near Memphis, Tennessee. Passengers were crowded on the deck above the boiler when it burst with lethal fury, throwing bodies into the air and water. Death came from impact, scalding, burning, and drowning. The blast killed more than 50 passengers and gravely injured many more as horrified onlookers watched helplessly from the shore. Newspapers spread graphic descriptions of the disaster, and public outcry demanded that Congress take action to prevent further calamities.

In the wake of the disaster, Congress granted federal funds to the Franklin Institute in Philadelphia to support a comprehensive study of steam-boiler explosions and their causes. The Institute issued a comprehensive report that was a tribute to the growing value of the engineering sciences in analyzing technological risks. The Institute's experiments identified the complex causes of boiler explosions, which were poorly understood. These included poor design with inferior materials, lack of understanding of thermal expansion, corrosion, and silt buildup. Investigators found that, while the boilers had safety valves, they often were not properly sized or positioned, were prone to failure, and were often deliberately disabled by steamboat captains racing to achieve maximum speed by operating at steam pressures beyond what the boilers could withstand. The Institute recommended rules for boiler design and construction as well as methods for periodic inspection and pressure testing to assure safe operation.

Translating the scientific findings into effective legislation, however, proved to be a thorny problem. The government of the young country had not previously regulated the safety of free enterprise; congress-men's objections to the potential invasion of private property rights prompted a protracted debate over the constitutionality of prescribing any restrictions on steamboat transport. The legislators preferred simply to hold those responsible for explosions accountable though civil or criminal liability, much as the law had previously dealt with stagecoach lines and other preindustrial enterprises. Many legislators thought that owners' enlightened self-interest and scientific advances were enough to curtail the risk. The bill addressing steamboat safety passed Congress in 1838, but it was stripped of rules governing boiler design and operations, and it contained only an ineffectual provision for inspections that left the inspectors susceptible to graft and corruption.

The congressional act of 1838 failed to curb the tragedies. Steamboat explosions increased over the following decade, and the death toll rose. By 1852, when rising public pressure forced Congress to enact enforceable legislation, steamboat accidents had claimed more than 7,000 victims, with 700 deaths occurring while the legislation was working its way through Congress. The 1852 act was precedent setting, for it created the Steamboat Inspection Service, the first federal agency charged with protecting the public from technological mishaps. The agency enforced design rules for boiler wall thickness, maximum steam pressure, safety valve location and more. The agency also performed mandatory annual boiler testing at 1½ times operating pressure and collected data on accidents and their causes. Inspectors exercised authority to conduct on board inspections, revoke pilot licenses, and levy penalties.

The effect of the 1852 legislation was dramatic, reducing steamboat deaths from more than a thousand the year before it was enacted to 45 the year after. Fatality rates remained low until the Civil War broke out, during which regulation became lax under chaotic wartime conditions. The most tragic accident was the April 1865 explosion aboard the steamboat *Sultana*, which was overcrowded with Union solders on their way home after release from Confederate prisoner of war camps. More than 1500 lost their lives as a result of the boiler explosion.

As peace returned, the number of deaths again dropped as the Steamboat Inspection Service gained control of the waterways. As railroads surpassed riverboats as the nation's primary means of transport later in the nineteenth century, the Steamboat Inspection Service's success became a model for protecting public safety: Congress established the Interstate Commerce Commission to curtail the dangers of the rapidly expanding rail industry. The stage was thus set for governmental regulation of other enterprises as the nation became increasingly industrialized.

* * * * *

Complementing the increasing involvement of governmental agencies, an abundance of standards for the design and operation of technology were promulgated by a variety of nongovernmental organizations: trade groups, professional societies and nonprofit testing laboratories. Some are discipline oriented and apply across many industries. For example, the National Fire Prevention Association maintains an extensive code that specifies grounding, circuit breakers and much more to assure that electrical devices will not result in fire or electrocution; the American Society of Mechanical Engineers sets standards for a wide variety of mechanical equipment, stipulating safety valves for high-pressure piping, minimum strengths on structures and much more. Other professional organizations specify standards for specific industries, such as those of the Society of Automotive Engineers.

A number of nonprofit businesses operate testing laboratories for certifying that products have met safety standards. In the United States, the best-known of these is Underwriters Laboratories, or UL, whose services now extend to many countries. A wide variety of products, businesses, insurers, and government agencies require UL certification for them to be sold or installed. The company is also deeply involved with the process of writing many of the standards that products must satisfy to obtain UL certification. Other organizations also play important roles in setting standards. For example, in the US, the Insurance Institute for Highway Safety performs crash testing on motor vehicles and publishes ratings of their crashworthiness.

A vast array of current safety standards has come from the private sector (i.e., they did not originate as government regulations). With few exceptions, the standards are written and kept current by committees consisting mainly of engineers drawn from industry, academia, and governmental bodies, with the majority of them coming from manufacturing firms. The standards they set are voluntary. However, they frequently gain the force of law since municipalities often adopt them in building codes, and state and federal agencies often incorporate them into their safety regulations.

Situations arise, however, in which the public may feel that such standards are not stringent enough, are not enforced, or do not adequately address important safety considerations. Consumer advocates and others may argue that the standards are suspect since many of the participants who write them are employed by the industries that they regulate. They argue that the members of standards committees may weigh the economic interest of their employers too heavily in relation to the public's safety. The standards also face another challenge. When new or innovative technology is involved, the standards may not be up to date. New hazards may appear that the standards do not adequately address in a timely manner.

The citizenry's concern that a technology is not safe enough and its demand for government action to provide more stringent safety regulations frequently swells in the aftermath of a disastrous accident, one that draws a great deal of attention in the news media because of its poignancy, large loss of life, or spectacular physical destruction. The more graphic the accident, the fewer fatalities are needed to attract public attention. The demand for safer technology is reflected in the political process as elected officials react to pressure and pass legislation to deal with the risk. Where the risks encountered are perceived as both new and substantial, a new federal agency may be established to regulate the technology. More often, the new legislation may reinforce or change the rules under which an existing agency operates. In other situations, reinforcement can be accomplished without passing legislation. An executive order from the White House or congressional pressure on an agency to revise its operations may serve as alternatives.

During the twentieth century, a number of safety-related federal agencies came into being, led by the US Food and Drug Administration, or FDA, in 1930. Later in the century, increased concern about technological risks led Congress to establish several more federal agencies directed toward controlling these. Prominent among them are the National Highway Traffic Safety Administration (NHTSA), created in 1966; the Occupational Safety and Health Administration (OSHA) in 1970; the Environmental Protection Agency (EPA) in 1970; the Consumer Product Safety Commission (CPSC) in 1972;

and the Nuclear Regulatory Commission (NRC) in 1974. In some cases, such as the NRC, the agency split off from a larger governmental department in order to separate safety regulation from the government's promotional functions.

A more recent separation took place following the Deepwater Horizon explosion; legislation created the autonomous Bureau of Safety and Environmental Enforcement (BSEE) to separate safety enforcement functions from federal leasing and revenue collection operations, all of which had been within the Minerals Management Service. In other areas, safety and promotion remain within the same government agency; this is the case, for example, with the Federal Aviation Administration (FAA). However, an independent agency, the National Transportation Safety Board (NTSB) investigates all transportation accidents and make recommendations to the FAA, the NHTSA, and other agencies that have the power to promulgate regulations.

Elected officials reflect public opinion in passing legislation to deal with technological risks. But only rarely do they or their staffs have the expertise to spell out the regulations in detail. Thus they write legislation in quite general terms, and the regulatory agencies must then translate law's intent into specific enforceable rules. Design engineers, operating personnel, enforcement officials and the public must adhere to these specifics.

The rulemaking process, by which the broad acts of Congress are translated into detailed regulations, requires opportunity for public comment, and in many cases hearings where the public can be heard. Frequently, such hearings are confrontational, with citizens expressing their fears and demanding government action even though the experts employed by regulatory agencies may consider those fears to be exaggerated or unjustified. Corporations, landowners, and other participants who have financial or personal interests that may be harmed by the proposed rules are likely to vigorously oppose the regulations under consideration, citing excessive cost, unintended consequences, or other shortcomings.

* * * * *

Regulatory agencies at the federal level arguably are among the more sophisticated practitioners of quantitative methods of risk assessment; foremost among these are cost-benefit and cost-effectiveness analysis. And yet the agencies find many obstacles to their efforts to apply these methods, as they confront the often-conflicting interests represented in public hearings and private lobbying efforts. Efforts to specify detailed safety rules bring together quantitative assessments, public perceptions, and economic interests to determine how safe is safe enough. Usually, the results are positive. However, negotiations invariably result in compromises between those who originally pressured elected officials for change and the corporations, government bodies, landowners, and others who have financial, aesthetic, and other vested interests that may run counter to the drive for stricter safety laws.

How do Congress and the regulatory bodies upon which they rely gauge what an adequate level of safety is? Sometimes laws are written in very concrete terms, but may result in requirements that prove to be unworkable. A prime example is the Delaney clause, named after the congressman who inserted it into a wider-ranging bill. The clause specifically directed that "the Secretary of the Food and Drug Administration shall not approve for use in food any chemical additive found to induce cancer in man, or, after tests, found it to induce cancer in animals."

There were at least two difficulties with the Delaney clause. First, by feeding mice or other laboratory animals very large doses of pesticides or other chemicals—some synthetic but others occurring naturally—found in food additives, the additives could be shown to be carcinogenic. Second, in 1958, when the Delany clause became law, residues only as small as one part in one thousand were measurable. But that has changed.

As analytical chemistry became more advanced, scientists were able to detect residues of one part per billion or less, and these minuscule amounts of the growing list of banned chemicals could be found in many foodstuffs. The Delaney zero-risk standard presented a conundrum for the FDA since the use of pesticides and herbicides had become a widespread agricultural practice; it was impossible to remove every last trace of these chemicals from the food supply. Moreover, the concentrations were so small that the agency considered the cancer risk to be negligible.

Congress made repeated exceptions to the law for specific chemicals, such as some widely used pesticides. The FDA, and later the EPA, circumvented the clause more generally by applying the *de minimus* principle, stating that contaminant concentrations of less than one part per million presented a negligible cancer risk, and therefore were acceptable. Later, in 1996, the law was repealed, and Congress replaced the zero-risk standard with a clause in the Food Quality Protection Act that stipulated a standard of "reasonable certainty" of no harm to consumers.

Legislation asserting that standards should provide "reasonable certainty of no harm to consumers" gives the regulatory agencies more latitude in determining levels of safety. For other risks, somewhat similar principles back off from the zero-risk stipulation to let an "as safe as possible" strategy hold sway. For example, in stipulating limits on the permitted exposure to radiation from radioactive materials, the ALARA principle ("As Low As Reasonably Achievable") has been adopted by federal regulators and international bodies. They interpret it as requiring that radiation exposure be reduced to the lowest level possible, but not to a level that would mean prohibitive costs.

While the foregoing approaches eliminate the dilemma imposed by legislation requiring zero-risk or absolute safety, they still leave those writing regulations with the quandary of what level of risk exposure corresponds to "reasonable certainty of no harm" or what is "reasonably achievable." Clearly, the more stringent the regulations are, the higher the costs incurred in implementing them. Those costs need to be weighed against the benefits in health or safety derived from more stringent regulations.

Given enough latitude, the regulatory agencies may be able to devise other more workable criteria. For some mature technologies, they may be able to evaluate the levels of safety of similar existing technologies that are already accepted by the public. We refer to these as "revealed preferences." Revealed preferences often provide a useful starting point, even though they sometimes are in conflict with what the public says it wants, which generally is referred to as its "expressed preferences." If revealed preference is the standard, the design of new products should be as safe as—and preferably substantially safer than—similar technologies that are already widely accepted.

Using such arguments, the Federal Aviation Administration certified two-engine jet airliners for transoceanic flights on the basis that new engines had been demonstrated to be so much more reliable than those on three- and four-engine aircraft of previous generations that there was actually less chance of the newer aircraft with two engines crashing than the older one with three or four engines. But is there enough experience with the new engine to prove that it is safer than the old one? The aircraft manufacturer has to convince the FAA of the engines' reliability. This requires extensive prototype testing on the ground, exhaustive flight-testing without passengers under extreme conditions, and then gaining flying experience with passengers, but within limited distances from airports where emergency landings can be made. Finally, transoceanic flights would then be approved.

* * * * *

If safety design decisions are to be rationalized beyond a gut feeling or comparison with earlier or existing products, the cost of safety improvements must be weighed against the benefit derived from them. In recent decades, cost-benefit analysis has become a frequently touted method for making safety decisions objectively. In principle, the method appears straightforward. But in practice, the cost-benefit calculation invariably becomes unpalatable—even morally repugnant—to political leaders and the public alike, because it requires putting a monetary value on a human life.

A number of different methods have been employed to estimate values of life, including court settlements in wrongful death suits and estimates of lost earnings resulting from premature death. The most acceptable approach, however, derives from the concept of the statistical value of life, or SVL. The idea behind the SVL is that, since government regulations reduce the risk of death, the proper measure of benefit is the willingness of a society to pay for the reduction in risk. Thus the SVL does not directly put a value on life. When properly understood, it circumvents many of the objections that otherwise arise.

The SVL is calculated by asking how much people are willing to pay for a small reduction in the risk of death, or conversely, how much of a pay increase they would demand to subject themselves to a small increase in risk. A basic example illustrates the concept:

Suppose two groups of painters of comparable skill are employed at a work site. One group works at ground level, while the other receives hazard pay for working at heights at which most falls would be fatal. Suppose one out of 1,000 painters is expected to suffer a fatal fall each year. To compensate for this added risk, the painters demand hazard pay of $5,000/year added to their wages. The statistical value of life is then

$$SVL = \$5,000/(1/1,000) = \$5,000,000.$$ Of course, for the estimate to be valid, it must be made for large groups of people, and also averaged over a wide variety of occupations.

Investigators may also use the increases in price that customers are willing to pay to reduce their risk in order to estimate the statistical value of life. For example, suppose an automaker offers an optional set of safety features that is estimated to reduce the risk of death in an accident by one chance in 10,000 over the life of the car. If consumers are willing to pay $800 for the safety features, then the statistical value of life is $SLV = \$800/(1/10,000) = \$8,000,000$.

Economists use both expressed preference and revealed preference techniques to make SVL estimates. Expressed preference entails interviewing large numbers of people who are exposed to risk. The economists employ carefully worded questions to obtain estimates of how much those questioned would be willing to pay for a small reduction in risk, and/or how much they would demand to accept a small increase in risk.

With revealed preference, the investigators examine wage or price data or other relevant records to determine how much citizens actually are willing to accept in decreased wages or increased prices to decrease their exposure to risk. Expressed preference surveys are less expensive to perform, but they are less reliable because they involve hypothetical situations and are more likely to be subject to bias.

Estimates of the statistical value of life have increased with time, due in good part to the inflation of both wages and prices. The estimates also vary significantly between studies. As a result, different regulatory bodies may use different values in their cost benefit analysis. Currently, the values used by federal agencies vary between five and ten million dollars. Once an SVL is established, the benefit of a regulation is the number of lives saved multiplied by the SVL.

To apply cost-benefit analysis, the cost of applying the regulation must also be estimated; this is often a prodigious task. It may range from a highway department estimating the cost of adding stoplights to a set of intersections to the National Highway Safety Administration determining the total increased price for all the autos sold in the US if it toughens the crash tests. In its simplest form, the criterion for promulgating a regulation is that the benefits be greater than the cost, or that the benefit to cost ratio

$$(number\ of\ lives\ saved) \times (statistical\ value\ of\ life) / cost$$

be greater than one.

If the issue is not whether to promulgate a regulation, but instead to choose which one of two regulations to promulgate, then the idea of cost effectiveness becomes relevant. The agency should choose the regulation that saves the greater number of lives relative to the costs incurred. Thus the regulator would use the above formula for each regulation and adopt the one with the larger value. Note that in applying cost effectiveness, the dollar amount used for the statistical value of life has no effect on the decision; it will not change between the two alternatives, and only the number of lives saved and the incurred costs will determine the choice of regulation.

Even when the value of loss of life or limb has been agreed upon, the application of cost-benefit analysis in the regulatory arena is a difficult and contentious process. Data must be assembled both on the number of lives likely to be saved by the proposed regulations and for the cost incurred. Both are difficult to obtain, especially for technology that is new and for which not many mishaps have taken place upon which to base fatality statistics. Even if the technology is more mature and those statistics are available, it may be difficult to estimate how many lives will be saved with the safer technology required by a new regulation. A great deal of uncertainty will remain until extensive experience has been gained with the safer technology.

Costs may be even more difficult to determine; most of the data on costs resides within the industries that manufacture and/or operate the technology under consideration. The regulatory personnel who do the analysis may not recognize some of those costs. Conversely, the industry targeted by the regulations may have reason to exaggerate the costs of regulation. The industry may be concerned that the new regulations will interfere with their business, hurt their competitive position, and cut into their profits.

The industries to be regulated may try to delay or defeat the proposed regulations by demanding rigorous cost-benefit analysis while withholding cost data necessary to complete the process. Or they may take a regulatory agency to court, arguing that the analysis was incomplete or biased. A number of regulations have been delayed or defeated with such tactics. On the other hand, by bringing pressure on their elected representatives, advocates for the regulation may force the agency to issue the regulation, bypassing the cost-benefit requirements.

Even if the regulatory agencies are able to resolve the foregoing issues, they must take a number of other factors into consideration in their cost-benefit determinations. They must, for example, factor injuries and disability as well as accidental death into their definitions of the statistical value of life. The agencies must also consider associated effects, such as changes in behavior that may result from regulation; some of these may lessen or totally negate a regulation's positive contribution to safety.

For example, if the installation of additional stoplights causes drivers to take more dangerous alternate routes that they deem to be faster, they may actually add to the risk of accidents. Likewise, if the Federal Aviation Administration adds safety restrictions that cause considerable flight delays or increases in ticket prices, the result may be that substantial numbers of passengers will decide to drive instead of flying. Since the fatality rate per mile traveled is far greater for motor vehicles than for commercial airliners, a large enough shift from flying to driving would actually increase the number of transportation fatalities.

Political appointees most often head the agencies within which risk analysts work, and the directions that the agencies take are often influenced by the leanings of the elected officials who appoint them. Some are likely to be more attuned to the public's perceptions of what the most serious risks are. They may bend to the pressures of public advocacy groups even though their demands may differ greatly from what statistics indicate are the predominant causes of accidental death, injury and disability. Other appointees may be more swayed by industrial influence, by the lobbyists who put pressure on them to go slow on regulation because of the dire consequences it may have on their business or on the growth of the economy in general. The regulatory process is deeply political and economic as well as scientific.

* * * * *

It was an automaker's nightmare. Ford's prized subcompact, the Pinto, seemed to be going up in flames. The Pinto was designed to be an American made alternative to low-cost Japanese and European small cars that were invading the US market in the late 1960s. Initially, the Pinto was a success, becoming the fastest-selling subcompact in the early 1970s. But then the auto was involved in a number of accidents in which it was rear-ended, causing its gas tank to explode and killing or severely injuring its occupants. Two high-profile lawsuits went to trial and gained great widespread television and newspaper coverage. As a result, the auto's reputation was damaged even though it met all of the standards of the NHTSA and the Society of Automotive Engineers.

Matters became worse when an investigative reporter gained access to a Ford document sent to NHTSA, requesting exemption from a safety requirement, and using cost-benefit analysis as a justification. The memo analyzed fuel tank ruptures, using $200,000 (roughly $1,000,000 in today's dollars) as the value of a life. This was the number provided by NHTSA. The memo concluded that, since millions of autos had been sold, only $11/vehicle ($55 in today's dollars) could be justified for fuel tank improvements based on the benefit of lives saved and injuries prevented. When the study's conclusions gained wide public attention as the result of a lawsuit filed on behalf of the victims of a fiery Pinto crash, an uproar ensued.

The public viewed the memo as evidence of corporate callousness and a lack of respect for the value of life. The cost-benefit memo dealt neither with the Pinto specifically, nor rear-end collisions in particular. Moreover, much statistical evidence indicated that the Pinto was no less safe than the imported subcompacts with which it competed. Further investigation revealed that Ford's crash tests had indicated a fuel tank problem with the Pinto, but that the automaker ignored it because of the cost of fixing it. The disclosure further fueled negative media coverage. A write-in campaign gained strength and pressured Ford to make a massive recall and NHTSA to strengthen its fuel system standards. The intense exposure of the Pinto's shortcomings caused the auto to lose favor with the public; eventually, Ford took it off the market.

The Ford Pinto example, punctuated by the public release of a cost benefit analysis, points to the integral part that safety plays in the processes by which technology is designed, built, operated, and maintained. Corporations must be concerned with the reputation of their products for safety, and the loss of revenue that results if they suffer from the adverse publicity that stems from safety hazards, whether or not they are exaggerated by media coverage. Likewise, they must bear the brunt of wrongful death and class action lawsuits, of government inquires, reprimands and fines if their work is found to be lacking. But manufacturers must also keep their costs under control. They worry that regulation will hurt their competitive position or increase the price of their products to levels that will cause customers to look for less costly alternatives; they may switch from electric to gas heating, for example, or travel by bus instead of by car.

Designers must make tradeoffs between safety cost, style, performance, and other factors. They must also consider tradeoffs between safety features. They may feel more comfortable employing cost effectiveness in choosing between safety features; in its most elementary form, cost effectiveness doesn't require assigning a value to a human life as does cost benefit analysis. It requires only that, if resources are to be spent on safety related design, the improvements should be those that save the most lives per dollar spent.

But even then, additional issues must be confronted. Among the most important, there are two approaches to increasing safety: decreasing the chances that an accident will happen, and lessening the adverse consequences of accidents when they occur. In determining a balance between the two, decision-makers must realize that one may save more lives while the other is perceived by the public to have a greater safety impact. The following hypothetical situation illustrates the point.

Suppose that an automaker found that he could save more lives by spending $100 per vehicle on brake improvements than he could in putting the same amount of money into added fuel tank protection. The difficulty would be that the benefit would show up only in the seldom-read annual statistics, which portray the reduced number of deaths per mile driven. That reduction would result from collisions avoided and probably amount to no more than a few percentage points. Meanwhile, the number of motorists who died as a result of fuel tank fires would remain unchanged, although much smaller than the number of lives saved by the improved brakes. Even if rare, those fatal fiery accidents would continue to gain full TV coverage and horrify the public.

While the brake improvement may indeed have saved more lives, it would do little to redeem the automaker's image as a manufacturer of safe automobiles. Thus to improve his marketing and profits, the manufacturer likely would focus on perceptions of risk rather than the number of lives saved and implement the safety feature that would be most visible to the public.

If an automaker decides to decrease the danger of fuel tank explosions, the engineers must figure how to reduce the risk. Once again, public perceptions matter. Consider the case of side-saddle fuel tanks. These tanks were located outside the steel frame of GM pickup trucks sold in the 70s and 80s. A number of side impact crashes caused the tanks to explode and resulted in fiery deaths that attracted widespread media attention.

Forced to improve fuel tank safety, engineers were faced with two possible paths: move the tanks within the frame of the vehicle, thus giving them added protection from impact, or leave the tanks where they were but reinforce them by adding internal bracing or increasing their thickness. Public perception matters to truck sales. So even if reinforcing the tanks would be less expensive and increase their safety as much as moving them inside the vehicle frame, the public would likely assume that a greater increase in safety would result from moving them inside the frame.

Such are the quandaries faced by corporate decision makers as they determine how much to expend on increasing the safety of their products and which safety measures should take priority. Whether in automobiles, appliances, airplanes, buildings or bridges, similar dilemmas are encountered as engineers grapple with the problem of the cost effectiveness of safety measures. Free market competition alone often does not provide adequate incentive to assure acceptable levels of safety. Product prices are important to customers, who cannot be expected to understand many of the subtle safety issues that underlie engineering design.

The regulations promulgated by governmental bodies are to some extent helpful in lessening the quandaries faced by product designers. They stipulate minimum safety standards, as do the regulations contained in the codes of good practice published by numerous professional societies and other nongovernmental agencies. Such regulations level the playing field in the sense that designs from competing companies must all meet the same standards. Moreover, the designers achieve some level of protection when accidents do occur simply by publicizing the fact that their products have met all of the required standards.

Regulations, however, cannot tell the designers in any detail how they should make their designs to assure that the requirements are satisfied. Doing so would muddy the relationship between regulators and the regulated. It would hinder the innovative talents of the design engineers. It also would make the regulators culpable if they specified details of a product's design, and that design turned out to be hazardously flawed. Thus, for example, the National Highway Transportation Safety Administration specifies speeds at which autos must survive crashes without injuring occupants, but it does not specify in any detail how those standards are to be met. The Federal Aviation Administration may specify jet engine reliability standards, but it does not detail how the engines are to be built. The agencies carry out tests to verify that their rules have been met, but they refrain from involvement in the details of the design process.

*****

Accidents that leave vivid images of death and destruction with the viewer are likely to be the focus of attention for TV reporters, and accidents that are memorable because of their bizarre nature are likely to gain extended media coverage. Likewise, illnesses seemingly caused by poorly understood technology, particularly industrial pollutants that have just been introduced, frequently draw extended media attention. Such events are prone to accentuate two phenomena that psychologists refer to as anchoring and availability. Anchoring is said to occur when a serious accident takes place, and the public judges other accidents in relationship to it. Availability is the propensity to feel that a particular class of accident occurs more frequently than it actually does if it readily comes to mind. These phenomena contribute to people's inordinate fixation on widely reported, sensational, or poorly understood causes of death; they seem to occur much more often than the more frequent mundane accidents that in fact result in far more deaths and injuries.

The emotional reaction to an accident vividly remembered may become contagious, leading to more media coverage, which subsequently results in more widespread and stronger public concern, which in turn causes additional increasingly scary headlines, becoming a self-sustaining chain reaction, progressing over many news cycles. With an increasingly panicky public reaction, pressure is put on public authorities, with calls for government action to decrease the risk. In some cases, public fear may be well-founded, and government action can lead to much-needed regulation. But in others, the fears may be exaggerated from sensationalistic or distorted coverage of the risk involved. Psychologists refer to the self-perpetuating cycle of media coverage and public concern as an "availability cascade." Even though scientific analysis may indicate that a risk is minor or nonexistent, pressure may prompt politicians in such situations to legislate ill-conceived regulations.

Other psychological phenomena also may magnify or diminish fears of particular technological risks. Arguably, whether the risk is voluntary and whether it is controllable are the most important factors in determining such fears. To some extent, driving a car is voluntary; if the risks appear too great, the driver can choose an alternate means of transport or cancel the trip. But if a gas line explodes under a family's house, that is far from voluntary and likely to be feared much more even though such explosions are much rarer than fatal motor vehicle accidents.

Even with a greater potential for risk, a driver is likely to make a trip by car anyway; not only is he taking the risk voluntarily, but equally important, he feels that he is in control. Chances are that he considers himself an above-average driver, capable of reacting safely to emergency situations. If uncomfortable with the risk, he can slow down or pull off the road; he can drive more or less cautiously, depending on his mood and the conditions on the highway. Moreover, if he were to take an alternate means of transportation, say a commercial airliner, he would likely feel uneasy; once he boarded the airliner, the risk would be involuntary and totally out of his control. Thus, even though air travel is much safer than driving a car, he would be more fearful of this experience.

Psychologist Paul Slovic and his colleagues have studied these and other risk perception phenomena extensively, including investigations that examine the various factors that heighten the public's fears. They find that the two most dominant ones are dread risks—those affecting large numbers of people—and unknown risks—those producing unknown consequence with delayed effects. Thus we proceed by separating ordinary accidents—those resulting in only one or a few fatalities that immediately follow an accident—from those where fear is magnified by dread and/or fear of the unknown. Chapters 8 and 9 deal with ordinary accidents. Chapters 10 and 11 discuss disasters that result in a large loss of life, falling largely into the dread category. Chapter 12 addresses the public's foreboding of long-term health consequences, particularly of those stemming from new or poorly understood technology.

Examining each of these broad categories of risk shows how methods for regulating risk employed by governmental bodies at the federal, state and local levels come together with the public's perceptions of those risks within a political process that is buffeted by vested interests—those of the industries that are to be regulated and of interest groups such as labor unions, public advocacy organizations, and the lawyers who represent those killed or injured in accidents. Risk assessment and regulation is indeed a complex and messy process that mixes political, economic and psychological factors in attempting to reach satisfactory compromises between competing interest groups to determine how safe is safe enough.

CHAPTER 8

Hazards on the Highway

THE 2013 TRAGEDY that took place on I-70 near Arriba, Colorado has been repeated all too often. The driver lost control of his sports utility vehicle (SUV). The SUV flipped and rolled three times before coming to rest. None of the eight occupants were wearing seat belts, and all were thrown from the vehicle as it rolled. Four died, all young people in their 20s and 30s. Others, including children, sustained serious injuries. It was unclear to police whether alcohol was a factor in causing the crash.

Looking into the causes behind this Colorado crash, however, demonstrates the difficulties encountered in reducing technological risks and understanding how industrialized nations determine what is "safe enough." Examining how the rollover risk came into being and the clashes between public health concerns, consumer behavior, and economic interests illuminates the interactions of technology and society.

Motor vehicle accidents are the largest cause of accidental death in the industrialized world, causing tens of thousands of deaths in the US alone each year.

In the 90s, SUV rollovers accounted for a substantial portion of those fatalities. The manufacture of SUVs originated in the Energy Policy Conservation Act of 1975, which set more lenient fuel economy standards for light trucks—including SUVs—than for passenger cars. From the late 1980s on, SUVs became increasingly popular for a variety of reasons, not least because they were perceived—and marketed—to be safer than ordinary cars. In some respects this was true, for they were larger and heavier than cars. The weight difference meant that, in a collision between a SUV and a passenger car, the occupants of the SUV suffered fewer injuries; the passenger car occupants were much worse off.

There was, however, a major safety drawback to the SUVs. They were top-heavy and much less stable than standard autos, which have a lower center of gravity. Statistics compiled by NHTSA showed that SUV occupants were dying in increasing numbers in rollover accidents. Prominent among these were single-vehicle accidents in which the SUV hit a curb or embankment, ran or slid off the road, or otherwise went out of control, and then flipped over. As SUV sales mushroomed, rollover fatalities exceeded 10,000 per year, with even greater numbers of severe injuries.

Product liability lawsuits have sometimes served as an alternative to government regulation in reducing the risks from dangerous products. The loss of profits that results from multiple lawsuits resulting in large settlements may force companies to redesign, discontinue, or recall the offending product. In addition, some suits may be fought out in high-profile court cases. These tend to attract widespread media coverage that alerts the public to the hazard, thus causing sales to plummet.

Neither of these mechanisms, however, was effective in reducing SUV rollover accidents. The auto companies settled the suits quietly, agreeing to higher compensation in exchange for the stipulation that the records must be sealed. This the companies could afford, because the profits that SUVs brought were huge; they were selling truck-chassis vehicles at luxury car prices.

For well over a decade, an extended battle centered about NHTSA's struggle to determine how—if at all—and in what form the federal government should intervene to lessen the danger of rollover accidents. Government recalls were out of the question, because the 1970s law under which NHTSA operated stipulated that an entire class of vehicles could not be recalled from the market because of safety-related characteristics.

The law allowed recalls only if a defect was found in particular models. That stipulation had been enacted to protect convertible sales; when convertibles rolled over, the consequences were frequently fatal. SUVs also fell under this law since—as a vehicle class—they were inherently unstable, and the law did not classify instability as a defect. In contrast, when a design flaw caused Firestone tires to fail and cause SUV rollovers, they were recalled. But the defective tires were responsible for only a small fraction of SUV rollovers.

With recalls out of the question, consumer advocate groups, insurance organizations, and trial lawyers argued for an alternative approach: mandating strict regulatory standards. The auto industry lobbied that such standards would make little difference because the accidents were caused primarily by road conditions and driver behavior, not vehicle design. The manufacturers also insisted that regulatory costs would greatly outweigh their benefits. If SUVs didn't meet the standards, they couldn't be recalled and fixed. The lack of stability was so deeply imbedded in the vehicles' structure that the SUVs couldn't be modified, and taking them off the road for the two or more years that it would take to totally redesign SUVs with a wider, more stable wheelbase would be economic catastrophe for the automakers.

As the controversy grew in the 1990s, the manufacturers had already invested hundreds of millions of dollars in the design and prototype testing of the next generation of SUVs, which were also unstable. The money invested would be lost if the regulations prevented the new models from being sold. Supported by congressmen from the states where manufacturing was centered, the manufacturers' associations were able to delay actions on SUV rollover regulation.

Early on, NHTSA, proposed a simple criterion, the "Static Stability Factor," as a measure of safety. The Static Stability Factor is defined simply as one half the width of the wheelbase divided by the height of the center of gravity: the larger the factor, the more stable the vehicle. The automakers argued that this index was not a reasonable measure of SUV stability; many other factors come into play, such as the suspension and braking systems.

Conversely, consumer advocates argued that the Static Stability Factor alone was not adequate. They argued that dynamic maneuver testing was needed to more accurately determine under what conditions the SUVs would flip. They likened these studies to crash testing with dummies used to determine front and side collision crashworthiness.

But the auto industry resisted; after some testing, NHTSA investigators concurred that comprehensive maneuver testing was not practicable because too many different scenarios caused the vehicle to go out of control and leave the roadway. NHTSA staff continued to insist that the Static Stability Factor should be the primary criterion; they argued that no matter what caused a SUV to run off a roadway, the Static Stability Factor was the primary determinant of rollover risk.

Facing continued argument, delaying tactics by the automobile industry, and pushback from congressmen who thought that regulation would hurt the local economies that they represented, NHTSA put aside efforts to impose regulatory standards. Instead, the agency proposed a rating system based on the Static Stability Factor that would allow buyers to compare the stability of vehicles as they shopped. The first attempt of the NHTSA administrators was to require the rating to be posted on the windshields of showroom vehicles. But a strong lobbying effort by the car dealers and their allies defeated this effort, leading to further delays.

Finally, in 2001 NHTSA promulgated a five-star rollover rating system based on the Static Stability Factor: the fewer stars, the more top-heavy the vehicle, and thus the more likely to flip in an accident. The least stable vehicles, rated at one star, are at four times the risk of rollover than the most stable, rated with five stars. The ratings were posted on NHTSA's website, and appeared in numerous magazine articles as well. Lobbying efforts, however, prevented the ratings from being required on new car windshields, where they would be most helpful to consumers.

The SUV rollover rate has declined with time. A study by the Insurance Institute of Highway Safety has attributed this to a number of factors. The introduction of a new generation of more stable SUVs has had a substantial impact. More important has been the introduction of electronic stability control, or ESC, to assist drivers in controlling their SUVs on slippery roads and in making sharp maneuvers. These became available on all SUVs purchased in the US The proportion of SUVs with ESC increased from less than ten percent in 2000 to roughly 85 percent in 2010. In 2012, it became mandatory that all passengers cars sold, including SUVs, be equipped with ESC.

Although the annual number of fatalities resulting from rollovers increased with the number of SUVs on the roads, the number of fatalities per SUV has deceased substantially to less than a third of what it was in 1990. But even now that ESC is required in all vehicles, the NHTSA predicts that more than 5,000 deaths will occur annually in SUV rollover accidents.

In addition to accident avoidance standards, such as the ESC requirement, NHTSA has attempted to increase the survivability of accidents by increasing requirements to prevent roof crushing. The 1973 standard that an overturned vehicle must be able to hold 1½ times its own weight was increased to three times its own weight in 2009. Still, the numerous complications faced by NHTSA in dealing with SUV rollovers illustrate some of the complexities encountered in attempting to reduce just one of the technological risks of modern day life.

* * * * *

The problems with SUV rollovers notwithstanding, the large number of motor vehicle fatalities each year is showing some encouraging trends. Over the last thirty years, the number of motor vehicle deaths has shown a downward trend, dropping from 50,000 to less than 35,000 per year even though both the number of vehicles on our highways and the number of miles driven each year is increasing. A more telling statistic for driving safety is the number of deaths per mile driven. Those numbers are decreasing much more rapidly than the number of traffic deaths overall. The likelihood of death per mile driven is now less than a third of what is was in 1975.

Many things contribute to this decrease in motor vehicle fatalities. Government agencies and industry alike strive to decrease the risks of highway travel. They work to reduce the frequency of accidents as well as the severity of those that do occur. They reduce death rates by increasing the crashworthiness of automobiles and by improving highways and other infrastructure as well. They employ traffic law enforcement, antilock brakes, limited access expressways and more to reduce the number of collisions. Likewise, they improve accident survivability with seat belts, airbags, and crashworthiness requirements on autos. Even with these favorable trends, however, motor vehicle accidents remain the leading cause of accidental death in the United States as well as abroad.

Congress gave NHTSA authority to establish motor vehicle safety standards as well as the responsibility for investigations and recalls to correct safety-related defects or noncompliance with federal safety standards. The statues governing NHTSA require that, in determining safety standards, the agency consider relevant available motor vehicle safety information to determine whether a proposed standard is reasonable, practicable, and appropriate for the particular type of motor vehicle, and to include in its deliberations the technological ability to achieve the goal of a standard, taking economic factors into consideration. Many of the legalities under which NHTSA operates came into play, and sometimes into conflict, in its dealings with the SUV rollover accidents. Along with other federal agencies, NHTSA has attempted to reduce some of the ambiguities encountered in safety regulation by employing methods of quantitative risk assessment.

Foremost among the assessment methods has been cost-benefit analysis, which NHTSA has applied to justify its rule-making. But each new proposal that it must evaluate using this approach may be fraught with difficulties. Even with the value of a statistical life determined by the methods discussed in the preceding chapter, the agency must still determine the number of lives that are likely to be saved by its action, as well as the cost of implementation. Collecting such numbers is expensive and time-consuming, for it requires extensive analysis of accident statistics. And if a new safety device is being proposed, the NHTSA must estimate how many lives it will save. Likewise, costs of implementing the regulation may require extensive analysis; the analysis may make use of figures provided by automobile manufacturers, figures that driver advocacy groups may dispute. Moreover, those opposing more stringent regulation may use the need for data as a delaying tactic.

Finally, even though value of life may have been specified for agency use, public pressure—in particular the nature of the risk—may increase that value. Consider, for example, that in 2007, Congress mandated that NHTSA take action to reduce backup accidents. Congressional action was prompted in part by the great deal of publicity that surrounded a case in which a father killed his toddler by backing into him. This risk had increased with the popularity of SUVs, which have very poor rear visibility. The result was a mandate to reduce such accidents by requiring automakers to install backup cameras in all new vehicles.

The law requiring backup cameras probably could not have survived if strict cost-benefit criteria had been applied, for the implementation costs may well have exceeded the benefit. NHTSA estimated that the cameras will prevent approximately one hundred deaths and several thousand injuries per year. However, since nearly all of those killed or injured would be under the age of five, NHTSA justified the backup camera rule by taking into account benefits that it could not quantify. That is, the agency assumed implicitly that society values the lives of young children much more than those of adults.

In some situations, NHTSA uses rulings to force changes in technology that are justified by cost-benefit analysis. Crashworthiness—so-called passive—standards may be difficult to evaluate, but crash avoidance—so-called active—standards present even larger problems. Invariably, little or no accident evidence is available to determine how well new crash avoidance regulations will work; even after they are mandated, evaluating how many accidents were avoided is difficult. Antilock brakes, for example, cannot be evaluated until collision rate statistics become available for comparable fleets of autos with and without the crash avoidance feature. In contrast, crash dummies and other testing procedures make laboratory evaluation of crashworthiness criteria less difficult to evaluate. The five-star rating system for crashworthiness for frontal and side collisions resulted from such tests.

* * * * *

In reducing fatalities, safety standards are important not only for automobiles, but also for the highways on which we drive them. And the responsibility for highways falls primarily to state Departments of Transportation, or DOTs, in charge of building and maintaining highway systems. At DOTs, and at local levels as well, cost effectiveness may be more relevant than cost-benefit analysis because state budgeting processes most often result in fixed amounts of money allocated for infrastructure improvement and maintenance. The DOTs must allocate the money among the three objectives: economic development, congestion reduction, and safety.

With a given budget for safety improvements, the issue is then how to expend available funds most effectively to achieve the greatest reduction in the number of lives lost and injuries incurred. Even when that goal has been specified, a number of thorny issues arise that officials must incorporate into their cost effectiveness calculations. Should resources be concentrated on improving "hot spots" where the greatest numbers of accidents have occurred? This might well involve putting median guardrails along particularly treacherous stretches of highways to prevent head-on collisions or straightening out curves to prevent run-off-the-road accidents. Or would the money be better spent on statewide efforts to provide rumble strips to alert drowsy drivers that they are crossing a center line, or signs to alert drivers that they are approaching a sharp curve? All such tradeoffs, as well as those between improvements and maintenance, must be figured into cost effectiveness calculations dealing with transportation infrastructure.

The strength of traffic laws and the rigor of their enforcement arguably are as important to the reduction of deaths and injuries from motor vehicle accidents as are the safe design of the vehicles we drive and of the highways on which we drive them. Returning to the example of rollover fatalities, many studies have shown human behavior, even more than equipment shortcomings, to be a primary cause of accidents. Most rollover fatalities result from single-vehicle accidents, many occurring at night with young men driving SUVs. Speeding and/or alcohol are factors in a majority of cases, and three of four rollovers victims are thrown from the vehicle, indicating that they were not wearing seat belts.

People drive too fast and take chances to save time, which in their subconscious cost-benefit analysis weighs more than the added accident risks that they are incurring. They drive while drowsy, or worse, under the influence of alcohol; they allow distractions—arguing with passengers, eating, fiddling with the radio, talking on cell phones, or texting and receiving messages. Thus driver behavior is prominent among the causes of accidents, and the authorities—state and local—are overtaxed in trying to reduce them. Budgets are shared between traffic safety and the prevention of property loss and violent crime, making it even more difficult to reduce the three major causes of traffic fatalities: speeding, driving under the influence of alcohol, and nonuse of seat belts.

Checking drivers for intoxication or for nonuse of seat belts only when stopped for other traffic law infractions catches just a small fraction of the violations. Randomly stopping motorists for blood tests of drivers and/or checks for seat belt use would be much more effective. But there are two problems. First is the great cost and need for additional personnel, which states and municipalities alike have been unwilling to finance to any great extent. Second, even if confined to times and locations where the number of violations are likely to be largest, a substantial majority of the stops would find no violations. Moreover, while the concern of detection would likely reduce alcohol use and seat belt nonuse, the inconvenience and embarrassment of blood tests—or even seat belt checks—brings about strong public pushback. Drivers' complaints of unwarranted interference, invasion of privacy, and other grievances result in political repercussions unwelcome to the law enforcement agencies; in response to such complaints, elected officials often move to cut back the programs.

Speeding is more easily detected; with speeding, the issue is likely to be whether traffic cameras or additional patrol cars are more effective. The presence of either tends to cause motorists to drive somewhat slower. However, if there is widespread violation, with traffic flowing five, ten or more miles above the speed limit, there is little the police can do other than arrest extreme cases—speeders who greatly exceed the commonly accepted level of violation. Public education programs, more severe punishment, and stringent driver's license requirements are likely to have positive but limited effects. Once again, human behavior and its tolerance to risk from commonly occurring situations thwart efforts to decease the deaths and injuries from motor vehicle accidents.

* * * * *

The difficulties inherent in developing safety regulations notwithstanding, automobiles continue to become safer as the reliability of brakes, steering and other components is enhanced, as safety features to improve crash avoidance are added, and as crashworthiness is improved. Over the last 30 years, seat belt use has saved more than 250,000 lives and frontal airbags nearly 30,000. Roads, too, have tended to become safer with better signs, more gentle curves, rumble strips and median barriers. Each time a step forward is taken, the accident statistics improve, but they don't reflect a lowering of death tolls to the extent that one might expect if driver behavior remained the same. The problem comes from behavioral change that falls under the heading of risk compensation.

Human nature, being what it is, often thwarts engineers' efforts to make driving safer. As we feel safer, our behavior is likely to become less cautious, and our decisions more prone to risk. We may drive a little faster knowing that we are belted in and that airbags will buffer the impact should an accident occur. We are more likely to venture out when conditions are bad—in fog, ice, snow, wind, or rain—knowing that we have antilock brakes and electronic stability control systems to protect us.

The phenomenon of risk compensation first came to light from the results of highway safety research, but it is also present in the interaction of humans and technology in many other arenas as well. The extent to which risk compensation negates attempts to improve highway safety is an issue of active debate. Some argue that the compensation can completely obliterate positive effects of safety features, and there is some data to back that up. Antilock brakes are a case in point.

NHTSA once considered antilock brakes likely to reduce rollover accidents, but years of crash statistics do not back this up. For reasons that researchers have yet to explain, passenger cars equipped with antilock brakes experienced an increased rollover rate. In addition, a controlled experiment with two sets of taxicabs in Munich, Germany, one with and the other without antilock brakes, showed that the taxis with antilock brakes experienced a higher accident rate than those without them! To whatever degree, risk compensation is just one of the many behavioral phenomena that make the reduction of technological risk a problem of psychology and economics just as much as it is of engineering.

Automotive design, road conditions, and driver behavior come together in determining the death toll from motor vehicle accidents that kill one or a few at a time; nevertheless, they dominate the accidental death statistics in industrialized societies. As automobiles have become safer, the preponderance of motor vehicle accidents now result from drivers' lapses in skill or judgment rather than from failures of steering, brakes, or tires.

As the foregoing examples indicate, behavioral shortcomings are considerably more difficult to correct than the design flaws or manufacturing defects in the automobiles themselves. Extensive effort has been expended in reducing the number of accidents caused by drinking and driving. Although some success has been achieved, the numbers of innocent people killed as a result of drunk drivers is many times what can be attributed to the shortfalls in automobile design or manufacture. Enforceable laws to curb unsafe behavior are much more difficult to sell politically than requiring failure-free engineering design.

Motor vehicle accidents are the largest single cause of accidental death in the industrialized world. But the number of fatalities stemming from other causes, when added together, is even larger. As we shall see in the following chapter, the interplay between product improvement and behavior modification determines the level of risk reduction achievable for many other causes of everyday accidents, such as the frequently occurring falls, fires, drowning, etc. that take only one or two lives at a time. The public takes little notice of these mishaps, even though they account for many more fatalities than the infrequent disasters that result in multiple deaths and spectacular destruction.

CHAPTER 9

Dangers at Home and at Work

Virginia graeme baker learned to swim when she was three years old. But that didn't help when she was playing in a hot tub connected to a friend's swimming pool. Underwater, she sat on the cover of a drain connected to the water filtration system and became entrapped by the drain's powerful suction. Her mother Nancy tried frantically to free the seven-year-old, but the pull of the filtration pump was too great, and her daughter died in her arms. The drain suction was so strong that it took two adults to pull the little girl free.

Flat drain covers similar to the one in that hot tub have caused other fatalities in swimming pools when swimmers, especially children, cover them completely, causing their bodies to be trapped and sometimes disemboweled by the overpowering suction. In other situations, the drains' powerful vacuum pull has caught victims' hair, causing dire consequences. The attempt to eliminate this danger is instructive in offering one example of the quest to assure that the wide variety of products that find daily use do not pose deadly threats.

The Consumer Product Safety Commission, or CPSC, had been tracking pool and spa drain accidents for years. Over a decade, suction entrapments caused 90 deaths and 63 injuries. While serious, these numbers were a small fraction of the nearly 300 drownings and thousands of water-related emergency room visits reported annually, many of them related to problematic consumer products.

The CPSC staff is spread thin, tracking many potential hazards, with those resulting in drowning accounting for only a fraction of the agency's wide-ranging responsibilities. With a heavy workload, congressional pressure for less rather than more regulation, equipment manufactures, resort owners and other affected parties lobbying against regulation, it is not surprising that CPSC had not imposed effective safety standards to eliminate the danger of suction entrapment.

All that changed with the years-long campaign launched by Nancy Baker to eliminate the hazard that had cost her daughter's life. During the campaign, a number of other children died in suction entrapment accidents that were well-publicized and increased national awareness of the problem. More importantly, Mrs. Baker had a powerful ally: Virginia Graeme was the granddaughter of former Secretary of State James Baker. To counter the propensity of safety standards initiated by federal agencies to get bogged down with delaying tactics by industrial lobbyists, extended hearing processes and often court challenges, Secretary Baker exerted influence with his widespread network of friends and former colleagues in Washington to pressure Congress to act directly. And it did. In December 2007, five years after the accident took place, the President signed the Virginia Graeme Baker Pool and Spa Safety Act into law.

The new law required pools and spas to meet basic safety standards, including the installation of anti-entrapment drain covers. Similar to other consumer product regulations, the legislation presented a number of challenges if its implementation were to be effective in eliminating life-threatening accidents. The law required all new equipment to meet the tougher standards, and it set a deadline for all public facilities to be retrofitted with equipment meeting the new safety requirements. The law stipulated steep fines for violations, but enforcing the deadline has been difficult, primarily because of the limited number of inspectors that the agency can commit to the task.

Investigative reporting indicated that, shortly before the law's deadline, very few of the new drains had been installed, with some owners unaware of the law's requirements. The second difficulty was that the law exempted residential facilities; that exemption included homes that were rented for months each year. The exemption was made to counter the many complaints during the rule-making process made by those who argued that requiring equipment replacements would be an undue economic burden on homeowners. That exemption greatly diminished the impact of the law; more than half of pool entrapments have occurred in pools associated with private residences. As a result, the risk will decrease only slowly, as old equipment is eventually replaced with new.

* * * * *

Swimming pool drain entrapment is only one of many causes of death that relate to faulty technology. While many hazards may account for fewer than 100 fatalities per year, they add up to the tens of thousands of accidental deaths each year that don't involve motor vehicles. In the United States, governmental responsibilities for determining adequate levels of safety are distributed between federal, state, and local authorities.

At the federal level, several agencies regulate risks, including regulation of medical devices by the Food and Drug Administration, of aircraft safety by the Federal Aviation Administration, and of workplace hazards by the Occupational Safety and Health Administration. Outside the workplace, the responsibility for safety falls most heavily on the Consumer Product Safety Commission, which deals with the hazards related to more than 15,000 consumer products and the many causes of death: falls, drowning, fire, electrocution, suffocation, and so on.

To carry out its responsibilities, the CPSC makes a large effort to track emergency-room visits and other accident-related data to determine how frequently specific products may be causing injury or death; it may also undertake investigations based on consumer complaints. A wide variety of products have undergone agency scrutiny, ranging from washing machines, lawnmowers, and off-road recreational vehicles to bed railings, children's toys, and baby equipment. The commission may take a number of actions to deal with the hazards that it identifies. It may order that the product be recalled and modifications made to eliminate the hazard. It may set safety standards for product design and manufacture; in more extreme cases, it may ban the product altogether. The latter was the case for drop-side baby cribs.

Drop-side cribs are those in which one or both sides slide down to give parents easier access to the baby. However, the slide mechanisms and latches are prone to failures that can lead to partial detachment of the drop-side rail from the crib. Such failures sometimes cause a V-like gap to form between the slide rail and the mattress. The danger is that a baby that gets its head caught in the gap may suffocate or be strangled. Since 2000, drop-side malfunctions have resulted in more than 30 deaths and nearly 1,000 injuries.

To deal with the growing crib problem, the CPSC recalled cribs found to have faulty designs, breaking slats, malfunctioning latches, or other hazardous shortcomings. In all, CPSC made twenty different recalls totaling more than 12 million cribs of many different designs. The problem was made more complicated by the way baby cribs are used. Typically, they are disassembled and stored, given away or sold for future reuse. Reassembly, however, creates problems such as loosened nuts or bolts, missing hardware, or lost instructions that may result in improper reassembly, greatly increasing the likelihood that a faulty drop-side will result in tragedy.

Finally, the CPSC concluded that recalls were not dealing with the hazards adequately; in 2011, the commission banned the sale of all drop-side cribs. Banning the manufacture and sale of a consumer product eliminates the hazard in the long term, but only after all of those already in use—and for many products it is in the millions—are worn out, thrown out, or otherwise retired from use. Banning drop-side cribs eliminated sales of new cribs immediately, but notifying resale shops and stopping the cribs from being recycled to new parents—often in degraded and more dangerous condition—was a more difficult challenge. Getting the word of the danger out to those currently using the cribs and to those who had them in storage with plans to reassemble for the next child or to give to friends, required a massive publicity effort.

* * * * *

Accidents often result from product failures, but they also result from a product being used or misused in ways that the designers hadn't anticipated. Thrill-seeking adults, for example, may subject recreational equipment to unreasonable abuse. Rollover accidents of all-terrain vehicles have caused many deaths. But if the commission regulated them to be more stable or limited to slower speeds, some users would complain that the lack of performance was impeding their recreational value.

Likewise, some owners may misuse appliances in ways that it would seem no reasonable person would think of. For example, if sticking your hand in a garbage disposal while it is running causes injury, is that reason to recall the product for redesign? A look at the warnings at the beginning of nearly any consumer product instruction book reveals a wide variety of misuses.

To clarify such situations, the CPSC makes a distinction between products intended for general use and those intended for children. More stringent criteria are applied for toys and other products intended for children's use. A child—particularly a toddler—cannot be expected to act with adult rationality. Thus the Commission requires toys to be tested for hazards by an accredited laboratory before they can go on sale: such hazards include swallowing small pieces, lead paint, sharp edges, flammability, the possibility of electrical shocks, among other hazards.

As is the case with other federal agencies, the CPSC must deal with stakeholders with diverse interests in setting its regulations. Consumer product manufacturers as a rule generally prefer voluntary standards that are reached with industrial consensus. With industry's emphasis on cost containment and noninterference by regulatory bodies, however, the voluntary standards may fall short from what consumers and their advocacy organizations demand.

Through its rulemaking authority, the CPSC has the power to promulgate mandatory regulations and to recall products as the agency identifies unanticipated hazards of a significant magnitude. But the process is often contentious; as consumer groups demand immediate action while the affected manufacturers resist the added costs and damage to their reputation that the regulations are likely to bring. In principle, cost-benefit analysis should lead to an objective balance that accounts for both views. However, as in the case of motor vehicle safety, in practice such analysis often proves controversial when applied to consumer products.

Often the agency must rely on industry for data in estimating the costs of implementing a regulation. It may be in industry's short-term interest to exaggerate those costs, skewing the results of the study in their favor. Consumer groups and the families of injured parties who are pushing for stronger regulation are likely to accuse industry of delaying tactics behind the long delays encountered in gathering the data on lives or injuries saved and on the costs incurred used for cost-benefit analysis. One case frequently cited is that of the rule-making on table saw safety. It took eight years from the time that the initial petitions of a hazard came to the CPSC to the time when the Commission posted its notice of proposed rule making. In the intervening time, table saw accidents had caused 4,000 amputations and many more emergency room visits.

The tolerance of unsafe behavior and the degree to which government agencies are allowed to impose standards to reduce particular technological risks varies greatly from culture to culture, even among industrialized nations. In the US, for example, a homeowner whose house has burned to the ground may expect a great deal of sympathy, even though he had been bypassing circuit breakers to enable use too many electrical appliances, smoking in bed, or knowingly engaging in other blatantly unsafe practices. In some cultures, such behavior would likely be criticized, the tragedy notwithstanding.

In some nations, the citizen's less tolerant attitude toward risky behavior reflects itself in more stringent regulation, such as the rules for home inspection and the requirements for repair. In the US, such regulations are seen by many as intrusions and are resisted as invasions of privacy. Only when a home is first built (or, in some states, when it sold) or undergoes major renovations that require a building permit do authorities have the right to enter and require that the structure adheres to safety mandates of building codes.

The greatest difference between the US and other developed countries with regard to safety arises from the citizenry's attitudes toward firearms. In most civilized nations, gun ownership is strictly limited, with stringent licensing requirements and tough safety standards to prevent accidental shootings. In contrast, the US places few, if any, restrictions on gun ownership, even for assault weapons! Furthermore, carrying concealed weapons in public places is allowed in many states in the US. Guns are the only consumer products that Congress has barred the Consumer Product Safety Commission from regulating, and the Center for Disease Control is no longer allowed to publish statistics on deaths and injuries caused by firearms.

Among industrialized nations, such restrictions on governmental regulation are unique to the gun culture of the US. Not surprisingly, the per capita death rates from gun-related homicides, suicides and accidents are much smaller in most other developed countries. The majority of the tens of thousands of deaths caused by firearms in the US each year are from homicides and suicides, and thus not classified as accidents. Nonetheless, hundreds of fatalities and thousands of injuries result each year from firearm accidents. Many of these are poignant, attracting media attention, for they often involve children playing with loaded guns and shooting themselves, playmates, or adults.

Both technological fixes to improve the safety of guns and regulations to curtail children's access to them could substantially reduce the number of accidents. Mandating that safety mechanisms and locks be incorporated into gun design, in addition to passing and enforcing laws to ensure that guns are stored unloaded, locked, and out of the reach of children, would provide an additional reduction in death and injury. In the US, however, the gun lobby has effectively blocked governmental efforts to place restrictions on gun design, ownership or use, even when the primary objective of the regulation has been to reduce the number of accidental firearm deaths suffered by children.

* * * * *

Occupational safety confronts challenges that are in some respects quite different from those in dealing with the risks of consumer, medical, or other products used by the public. While consumers may need to weigh the increased prices that reducing risks may bring, often employees must contemplate the loss of their livelihood if the cost of increased occupational safety causes the company for which they work to become unprofitable.

Likewise, consumer product safety issues are more likely to affect large and diverse segments of the population, thus drawing media attention and political pressure for change. Conversely, many of the varied occupation health and safety risks may have dire consequences, but they may be confined to the small and often underprivileged groups of workers employed in a particular industry. It follows that occupational safety issues often receive scant media coverage, and little political pressure for change will be brought to bear.

Before workers' compensation was initiated a century ago, there was little cost to industry in ignoring safety problems. That changed with workers' compensation. While varying significantly from state to state, in general workers' compensation requires factory owners to pay an insurance premium to cover the cost of injuries, as well as for life insurance for cases in which workers are killed on the job. The size of the premium increases if a firm's record is poor with respect to the numbers and severity of injuries incurred. Thus an incentive is created for companies to improve the safety of their workplace—not doing so would have adverse financial consequences.

Nevertheless, workers' compensation alone proved to be inadequate, and in 1971 Congress established the Occupational Safety and Health Administration (OSHA) to reduce the number of deaths and injuries resulting from workplace accidents. Similar to the difficulties faced by other regulatory agencies, OSHA's efforts to improve safety are often contentious, involving issues at the interface of science and politics.

Industrialists prefer partnership relationships in which OSHA works with industry in developing standards, helps in achieving compliance, and participates in worker safety-training programs. But companies are less favorably inclined toward mandatory regulations imposed by the agency, particularly when they are expensive to implement. And when the agency shifts from assistance to enforcement, levying fines for violations, the regulated firms frequently lobby to limit the agency's scope of operations.

In contrast, unions, lawyers, and other advocates of increased worker safety argue for more stringent regulations and more frequent inspections to ensure that safety standards are enforced. They argue that, in trying to shift the emphasis from equipment safety to worker safety training, industry is trying to shift the responsibility for accidents from poorly designed or faulty equipment to the errors that workers may make using that equipment.

Cost-benefit analysis again comes into play, with employers frequently emphasizing the costs and interference imposed by regulations that may negatively impact their "bottom line." The workers' advocates stress the dangers inherent in the workplace and sometimes accuse the manufacturers of downplaying the number and seriousness of the injuries incurred in their plants.

Still, when employees' livelihoods are at stake, when strict enforcement of safety standards might cause employers to shut down a factory or plant, or to move operations elsewhere, the workers sometimes promote lax standards. Consider, for example, the situation reported in West Virginia, where 29 coal miners were killed in an explosion at the Big Branch Mine. Journalists from *The New York Times* found that, when local people saw the federal investigators coming to inspect the mine in their distinctive white SUVs, they would alert the mine operators, giving them time to clean up or hide some of the most egregious safety violations. Convenience store clerks would call in their warning while the federal inspectors stopped to buy coffee, and coal-truck drivers would use their CB radios to let the operators know the inspectors were headed for the mine.

The West Virginia community, in fact, was doing a crude cost-benefit evaluation. It had concluded that the loss of jobs by possible closing of the mine for its gross safety violations would be a cost too great to bear. In order to maintain its economic base, the community was willing to tolerate the number of lives lost and injuries sustained by the miners as a result of mine operators' disregard for safety measures.

* * * * *

The difficulties in assuring compliance with workplace safety standards are compounded when many small companies are involved. OSHA inspectors need to visit many different sites to cover many small companies, making inspections less frequent than for large centralized facilities with many employees. The dangers to workers of the more than 13,000 commercial grain elevator operators across the country that come under OSHA regulation offer a noteworthy example. The dangers of working in these elevators and bins are many, but the most horrific is suffocation resulting from a phenomenon known as grain drowning.

Grain drowning occurs when a worker is caught in a silo or bin from which grain is being withdrawn from below. The flowing grain creates suction, much like water going down a drain, pulling the victim under, and covering him with grain within a few seconds. Grain drowning also may occur while workers are "walking down" the grain, i.e., loosening grain that has become stuck to the bin walls. Large masses of grain may suddenly come loose, creating an avalanche that may bury the workers alive.

Grain drowning fatalities have been a persistent cause of death, but they rarely attract more than fleeting local media attention. An exception occurred in 2010, a year in which authorities recorded 26 grain drowning fatalities. Investigative reporting by national media publicized the details of a horrific accident in which two teenaged workers died, and a third narrowly escaped, while walking down grain in a Mount Carroll, Illinois elevator. The three boys were using shovels and picks to knock down corn that had crusted to the sides of the bin. Other employees were withdrawing grain from the bottom of the bin, and when they opened a second outlet to increase the rate of withdrawal, the 14-year-old boy was sucked into the flowing grain; one of his friends also died attempting to save him.

Reporters from National Public Radio found that the storage facility operators were violating many OSHA rules: the boys had no safety training and they were unaware of the need to wear safety harnesses, which were gathering dust in an adjoining building. Furthermore, one of the boys was underage for work in such a facility. More important, grain should not have been withdrawn from the bin while workers were inside it.

Operators willing to take a chance that no serious accidents will occur have little incentive to enforce safety standards that may add even minimal cost to their operations. Inspections are rare, and even repeated fines for violations are small enough that often bin owners consider them just another cost of doing business. Even when a serious accident leads to death or disabling injury, the OSHA-imposed fines are only a fraction of the value of life used in cost-benefit analyses. Moreover, since the fines can be fought in court, the agency frequently settles for a much lower amount in order to avoid tying up OSHA's scarce resources in lengthy litigation. In the Mount Carroll case, negotiations resulted in the initial fine of $618,000, but the fine was settled for well under half that amount.

Many thousands of silos and bins are located on smaller farms or are operated by businesses with fewer than 10 employees; these have been exempted from OSHA regulation. So also are children working with parents or other family members, under the assumption that parents will take great precautions to protect their children. But such work is dangerous, and substantially more grain elevator fatalities take place at these unregulated locations than at those covered by federal safety standards.

Even a complete tally of grain elevator deaths would amount to fewer than 100 per year, a small fraction of the more than 4,000 workplace deaths each year in the US Agricultural accidents, like those in factories, mines and elsewhere, present similar challenges to

OSHA. There are too few inspectors, and the fines are too small to greatly reduce the number of fatalities. Often there is complicity from those who cooperate with their employers in evading safety standards that would even slightly impede productivity, due to their fear of losing their livelihood. That fear outweighs the fear of the risks to which they may be subjecting themselves. In a way, it is a crude cost benefit comparison employed by the workers themselves that discounts what they presume is the small chance of death or injury relative to the greater chance of being laid off or fired.

* * * * *

Workplace health, as well as safety, comes under the jurisdiction of OSHA, and in many instances it presents even more of a challenge. Death or injury from falls, suffocation, electrocution or dropped objects are immediate, with cause and effect indisputable. Occupationally caused illness often is more insidious, for cause and effect may be more difficult to prove when it comes from chronic skin contact with dangerous chemicals or from breathing contaminated air. Early symptoms may be ambiguous, and debilitating illness or death may come months or years later. Yet, more disabilities and premature deaths can be traced to such chemical exposure than to work place accidents.

Assessing the occupational risks from the use of toxic or carcinogenic chemicals presents a set of challenges different from the assessment of risk for the exposure of much of the population to lead, mercury, and asbestos. Likewise, there is a difference in assessing occupational health hazards to those directly exposed and the public at large. A hazard caused by a workplace chemical that affects only those who work in particular industries where it is used usually does not get much media attention, and the public in general remains for the most part oblivious to a hazard that doesn't affect them.

Pressure for reform is further impeded because affected workers often are economically disadvantaged and lack the influence to put pressure on political leaders. Moreover, like other workers, they may be careful about demanding healthier working conditions for fear of being fired if they complain, or that the plant will be closed or relocated in another country where health and safety regulations are less stringent or nonexistent.

Some OSHA regulations do address the hazards arising from dangerous chemicals. They specify safety gloves for limiting skin exposure to dangerous chemicals and require that respirators be worn where the air contains unsafe levels of contaminants. But often such general criteria do not adequately deal with the hazards presented by specific chemicals. And the challenge is daunting. Approximately 85,000 chemicals are used by industries in the US But the EPA has tested only 200 of them; since its establishment over 40 years ago, OSHA has specifically restricted the use of only sixteen, among them lead, asbestos, and arsenic.

OSHA has been criticized frequently for placing too much emphasis on safety and not enough on the health hazards caused by workplace chemicals. In truth, the same problems that limit OSHA's ability to decrease the number of workplace accidents are exacerbated in its efforts to reduce workplace health hazards. There are far too few knowledgeable inspectors to make frequent on-site visits, and the fines that may be issued are far too small to have a detrimental effect on employers. Proving that chemical exposure caused a specific health effect is more complicated than showing that a worker was injured or killed because he fell off a ladder, crushed by a machine, or electrocuted by faulty wiring.

Health inspectors must be more knowledgeable than those inspecting for safety hazards, and each inspection requires much more time. A further complication is that the fines OSHA may levy are much lower for health than for safety violations. Finally, and more importantly, because health effects may be complicated by workers' personal habits, and because they may not appear until months or years following chemical exposure, the fines and other mandates of OSHA are more likely to be successfully contested in court than are penalties for safety violations.

Without question, occupational exposure to toxic substances is a huge health problem. Estimates are that occupational exposures cause more than 40,000 premature deaths each year and 200,000 disabilities in the US. Yet the wider public pays little attention to these hazards relative to the other technological risks that it may fear. There are exceptions, however, when media coverage brings attention to a particularly shocking situation, such as the one recently exposed in the furniture industry.

* * * * *

N-propyl bromide (nPB) is used by thousands of workers in dry cleaners and body shops, as well as in high-tech electronics manufacturing. Its use greatly increased when it was substituted for another chemical—1,1,1-trichloroethane (TCA)—which was banned by the United States and other countries because it was causing damage to the ozone layer. But nPB was substituted for TCA in dry cleaning and body shops even though, for more than a decade, government officials and medical researchers have known that inhalation of low doses of nPB over a long time causes neurological damage and infertility. The increased use of nPB is just one of a number of situations where a government agency banned one chemical, and industry—left to its own devices—found a substitute that turned out to be yet more dangerous.

Investigative reporting by Ian Urbina of *The New York Times* exposed the hazards of nPB at the Royal Comfort Seating factory in North Carolina. The yellowish nPB mist permeated the workplace where the glue was applied to polyurethane foam because nPB is cheap, fast-drying and strong. Within the first days of workplace exposure, the fumes caused workers to suffer headaches and dizziness; within weeks, they began to suffer other symptoms that became progressively worse, particularly severe back and leg pain. Some workers had to stand on cushions brought from home to relieve intolerable pain in their feet and legs.

As the toxicological damage increases, the pain becomes intolerable, and the workers could not stand or sit for the extended periods of time required in their work. Eventually, they would lose control of their hands to the point where they were unable to dress or walk without assistance. Inevitably, the disease caused those exposed to be absent from work for longer and longer times and caused them to lose the ability to perform their assigned tasks with the speed and accuracy demanded by their employer. The company then fired them. Over a five-year period, federal authorities have documented severe nPB nerve damage to more than 140 of the cushion factory workers.

Attempts by OSHA to rectify the situation at the Royal Comfort Seating factory have been largely ineffective, for the inspections are too infrequent and the fines imposed inconsequential. OSHA demanded improved ventilation to reduce the fume concentration, but the company then turned the fans off to save money; when OSHA required the employees to wear safety masks, the company provided only cheap, ineffective masks. Likewise, getting the chemical banned from use altogether would be difficult, for the procedures needed are lengthy and expensive, and the budgets of OSHA and the EPA are hardly adequate to deal swiftly with the potential hazards of thousands of industrial chemicals.

The situation at the furniture factory illustrates the difficulties in reducing the risks of occupational exposure to toxic or carcinogenic substances, and how this differs from environmental risks that face the public at large who may be alerted through the media and become vocal advocates for change. The workers at the furniture factory were unorganized; they had no union to protect them. They were also desperate for employment, and they knew that the company would have no difficulty in finding others to replace them. They also realized that if the government were successful in imposing adequate health requirements, the cost of production might increase and cause the manufacturer to close the plant and move offshore to a country where labor was less expensive and health and safety regulations even more lax.

* * * * *

Governmental bodies expressing the will of citizens have a number of tools that they can apply to reduce the numbers of fatalities from the everyday accidents that arise from a variety of causes at home and in the workplace. None of the remedies are perfect; they are often difficult to apply fairly. For mass produced products, governmental power to set standards, mandate safety improvements and force recalls is the most straightforward means for dealing with faulty design or defective manufacture; success requires neither changes in human behavior nor the employment of huge numbers of inspectors. Regulations that require changes in attitude or behavior of citizens, employers, or employees are more difficult to apply successfully, particularly those that deal with risks that are undertaken voluntarily, where the risk takers feel that they are in control, or where the risks are so commonplace that they are taken without second thought.

The nature of those risks and the ways in which society attempts to curb them through regulation, law enforcement and education varies according to the venues in which they occur. They differ between public places, the workplace, and the home. Consider, for example, the risks of fire. Smoking in bed, misuse of alcohol, improper uses of electrical appliances and other human shortcomings are major contributors to fatal fires.

Certainly, smoke detectors, fire-resistant building materials, sprinkler systems, and other technical advances can greatly reduce the susceptibility of buildings to fire. The technology of building construction has improved much more dramatically, however, than has human behavior. Likewise, substantial strides have taken place in the design of pleasure boats to make them nearly unsinkable. But failure to wear life jackets, to remain sober or to stay on shore in bad weather often negates the improvements made in ship design.

When risk reduction requires not simply the improved design of equipment, but an assurance that it is being maintained and used in a safe manner, law enforcement of one form or another becomes involved, and privacy becomes an issue. It is least problematic in public domains, on our highways and in public buildings, where inspection is most easily accomplished. It is more difficult in the workplace, where the law must grant inspectors permission to enter; to be effective, they must be allowed to arrive unannounced. But that is often difficult.

Dangers in the home are the most difficult to eliminate. Building codes and permits may require smoke detectors, wiring that meets electrical codes and more. But once the building is occupied, wiring may be modified, smoke detector batteries not replaced, and circuit breakers bypassed. Furthermore, government officials are not allowed to enter—unless there is a search warrant for criminal activity—and even if they could, agencies with safety enforcement duties would unlikely have the personnel available to inspect more than a minuscule number of the places for which they are responsible.

Reducing the chances of accidents arising from bad design and faulty construction is engineering's responsibility. It is necessary to build products that are forgiving of inappropriate use or abuse as well as of improper installation or maintenance. But steep reductions in fatality rates are difficult to achieve, for they require the political will and social consensus to curb the behavior that contributes greatly to the car wrecks, boating accidents, fires, and many other everyday accidents that take thousands of lives each year.

Public risk tolerance varies greatly with the nature of accidents, the larger one-off event provoking tougher demands on technologies that actually pose less risk potential overall. Greater safety is demanded of airlines than of automobile travel. Higher accidental death rates are tolerated from backyard swimming pools than from chemical storage facilities, and greater fire protection is demanded in public buildings than in individual homes.

All such comparisons reflect the complexity of deeply held attitudes toward risk. To examine the issues involved, technological accidents were earlier divided into three rough and sometimes overlapping categories: ordinary accidents, disasters, and health hazards. The ordinary accidents, most frequently motor vehicle accidents, that take one—or at most a few—lives at a time account for the preponderance of deaths and injuries. Disastrous events, which in fact take many fewer lives, and concern with the public heath risks are the focus of the chapters that follow.

CHAPTER 10

Man-Made Disasters—the Problem of Probability

Two BOEING 747 jumbo jets, a Pan Am flight from New York and a KLM aircraft originating in Amsterdam, headed toward the Canary Islands on March 27, 1977, where they were scheduled to land at Gran Canaria Airport. Located in the Atlantic several hundred miles off the northwest coast of Morocco, the Spanish outpost is a popular point for vacationers to arrive by air and transfer onto cruise ships. But flight plans changed abruptly when a separatist movement exploded a bomb in the airport restaurant and called with a threat that another would follow. As a precaution, air traffic control diverted the two 747s and a number of other airliners to the smaller Tenerife Airport on a nearby island. So many aircraft landed at Tenerife that several of them, including the two 747s, had to be parked on the only taxiway.

Two hours later, the bomb threat had been resolved and the aircraft cleared to proceed to Gran Canaria. However, with the taxiway blocked, takeoffs had to be carried out using a tricky procedure in which the KLM 747 would taxi down the runway, followed by the Pan Am aircraft. Near the runway's end, the Pan Am plane would pull off the runway into a remaining parking spot. The KLM plane would then make a U-turn and take off, to be followed at a safe distance by the Pan Am flight.

Unfortunately, a dense fog set in as the operation began, and the control tower could not see the two aircraft, nor could the two pilots see each other. Worse yet, the air traffic controller could determine the aircrafts' positions only by radio since the airport had no ground radar. Communications became muddled, in part because messages from the two aircraft interfered, and disaster ensued. The KLM pilot attempted to take off before the other 747 could exit the runway. Not high enough off the ground, the KLM 747's engines gashed into the Pan Am fuselage and set it afire. The KLM pilot was able to keep his plane airborne for only a few hundred yards before it too crashed in a fiery explosion. The horrific collision killed all the KLM passengers and crew, and only 61 from the Pan Am 747 survived. The total death toll of 538 made it the most deadly disaster in more than a half century of commercial jet travel.

* * * * *

Commercial aviation has not been alone in spawning disastrous accidents. Widespread air travel came into being only in the mid-twentieth century. For more than a century before, railroads had been the primary means of long-distance land travel, and ships had crossed the oceans for much longer. In earlier days, frequent rail and ship disasters occurred, often taking a hundred or more lives in a single accident. As time passed, transportation technology became safer, and disastrous accidents became less frequent. Nevertheless, combinations of unusual circumstances still sometimes come together with disastrous results. As with air travel, the lessons learned from such events frequently lead to safety improvements that bring about further decreases both in the probability and severity of future disasters.

The deadliest US train wreck in decades occurred on September 22, 1993, as the result of a combination of circumstances that bordered on the bizarre. Some distance from Mobile, Alabama, a towboat was pushing a barge on the Mobile River. Night and dense fog confused the inexperienced towboat pilot, who had not been adequately trained in the use of radar. He turned into an unnavigable channel and rammed the barge into the Big Bayou Canot Bridge, thinking that he had run the barge aground.

The bridge structure was particularly susceptible to impact because of its swing bridge design, in which the mid-river section could rotate, allowing smaller boats to pass by. The impact forced the bridge out of alignment by about three feet and severely kinked the railroad tracks. It was unfortunate that the welded track didn't just break: an electronic track-checking circuit in operation at the time would then have caused stop signals to appear on the approaches to the bridge. Without that warning, Amtrak's Sunset Limited with 220 aboard sped through the fog and onto the bridge.

The train derailed, smashing its three locomotives into the bridge superstructure, with the first slamming explosively into the canal bank. The impact ruptured the fuel tanks of all three engines, creating a massive diesel spill and fire. The second and third locomotives, along with four cars, including two of the six passenger cars, plunged into the river. The results were devastating, killing 47—many by drowning—and injuring 103. It was the worst train wreck in Amtrak's history.

* * * * *

Air travel surpassed ocean liners years ago as the dominant means of transoceanic travel. But for shorter trips, ferries play vital roles in both developing and developed countries, and cruise liners have grown in popularity for vacation and educational travel. Ferry accidents are a frequent cause of large loss of life in the developing world. Though much less so in industrialized countries, rare but devastating accidents have taken place even there. One of the worst was the capsizing of the *Herald of Free Enterprise* off the coast of Belgium.

The *MS Herald of Free Enterprise* was a modern eight-deck ferry owned by Townsend Thoresen. Its design allowed quick loading and unloading of cars, buses and trucks to speed the vessel's tightly scheduled trips across the English Channel between England and Belgium. However, on the night of March 6, 1987, a design shortcoming combined with human error caused the doors in the ship's bow, used for loading and unloading the vehicles, to be left open. The crewmember in charge of door closings was asleep in his bunk, while the First Mate left the deck to go to the bridge without checking to see that the task had been accomplished. The captain could not see the doors from the bridge, and the ship had a major design flaw: there was no electronic signaling system to indicate that the doors were still open.

As the ferry headed out of port, water rushed in and quickly filled the length of the cargo deck. The inrush caused the ship to lose stability, to list, and then capsize within a mile of the shore. Fortunately, it veered to the side before capsizing and came to rest in the shallow water above a sandbar. The flooding caused all electrical circuitry to fail, throwing passengers and crew into darkness and leaving them trapped inside in the frigid water. The ferry's plight was spotted almost immediately, and rescue soon arrived, both with a helicopter and Belgian Navy ships that had been maneuvering nearby. A majority of the 539 occupants on board were rescued, but 193 died, many from hypothermia. Tragic as the loss of life was, it could have been much worse had the ship made it into deeper water before sinking, or had rescuers been slower to arrive.

* * * * *

Since the days of Mississippi steamboat explosions nearly two centuries ago, the disastrous sinking of ships, wrecking of trains and crashing of planes have frightened and fascinated the public. But other technologies—those that do not deal with transporting people in large numbers—have also led to accidents of calamitous proportions: fires, explosions, and structural collapse.

Fortunately, as in the case of transportation disasters, lessons are learned, public pressure is brought to bear, and practices are modified to reduce the risk of future accidents. Burning or collapse of structures housing large numbers of people is a case in point. Historically, such structural disasters have led to large losses of life, but they have also resulted in many of the safety measures in place today. In the United States, a disastrous factory fire in New York was such a turning point.

The Triangle Shirtwaist Factory fire took place in New York on March 25, 1911. It caused 146 of the young women working in the garment factory to lose their lives. At the time, there were virtually no enforceable safety standards in place, and the women were trapped in the eighth, ninth, and tenth floors of the building. The factory's management had followed the practice, common at the time, of blocking stairwells and exits to prevent theft and unauthorized entry. As a result, the occupants could not escape. They died of smoke inhalation or jumped to their deaths.

The Triangle Shirtwaist tragedy was vital in bringing about legislation that mandated more stringent safety standards. It also fostered the growth of the Ladies' Garment Workers' Union, which fought to improve the sweatshop working conditions. Today, conditions similar to those in the Shirtwaist factory still exist in developing countries, and tragedies take place all too frequently, as witnessed by two recent fires. In both fires, building safety standards were ignored or nonexistent and escape paths blocked to avoid theft. Only a few months apart in 2012, at least 289 workers died as fire engulfed a garment factory in Pakistan, and 112 lost their lives in a similar factory fire in Bangladesh. And yet again, when the Bangladesh garment factory building collapsed, as discussed in chapter 6, more than a thousand workers lost their lives.

Building fires and collapses have become much less frequent in the industrialized world and the death tolls less severe. Nevertheless, vigilance must be maintained even with modern safety standards in place, for lapses can still occur. A flaming example, the MGM Grand Hotel in Las Vegas, illustrates the point. On the morning of November 21, 1980, roughly 5,000 people were in the 23-story hotel and casino when a fire broke out in one of the building's restaurants and quickly spread into the hotel tower. Elevator shafts and open stairwells allowed smoke to quickly engulf the upper floors of the hotel. More than 1000 guests fled to the building's roof and were eventually rescued by helicopters from a nearby Air Force base. Nevertheless, 85 people died, primarily from smoke inhalation, and more than 650 were injured. It was one of the most deadly fires in recent US history, and demonstrated the need for continuing vigilance in places where large numbers of people gather.

* * * * *

Explosions, fires, or spills of the many flammable and highly toxic materials upon which industrialized economies depend also have led to the potential for disasters. Spills of sludge or liquids may cause grave environmental damage, but gases and volatile liquids cause greater immediate threat to human life. Such releases may occur from industrial processing plants or during the transport of such materials, whether by pipeline, tanker truck or rail. The loss of life in the industrialized West has been small relative to the thousands of deaths resulting from the Bhopal India catastrophe discussed in chapter 6, ranking it as the most deadly industrial accident in modern times. Still, a number of accidents in developed countries point to the need for never-ending vigilance.

For the most part, fatalities from explosions, fires, and toxic releases from industrial facilities in the developed world have tended to be primarily those of workers in the plants. The combination of restricting population in the immediate vicinity of the potentially dangerous plants and, more importantly, the execution of plans for rapid evacuation from areas where plumes of toxic chemical may endanger public health have greatly reduced fatalities outside the plants. Nonetheless, costs for cleanups and economic dislocations are often astronomical.

Freight transport presents different challenges; although the inventories of dangerous materials being transported may be much less than those stored at plant sites, the routes of trucks and trains often pass though highly populated areas, and the plans for evacuation are not likely to be nearly as well developed as those for areas adjoining industrial facilities. Moreover, people engaged in travel, occupying structures or employed at potentially dangerous worksites are there for pleasure or economic gain. But people who suffer from an explosion on a rail line or a toxic chemical release from a barge are more often innocent bystanders, put at risk only by where they live. Thus there is all the more reason why preventive measures and emergency planning are vital in reducing the public's risk from chemical transport accidents.

The challenges are well illustrated in Mississauga, Canada where, on November 10, 1979, a failed undercarriage caused derailment of more than half of the 106 cars on a freight train carrying explosive and poisonous chemicals. Tank cars filled with propane, chlorine, caustic soda and other chemicals ruptured, releasing the contents to the atmosphere and to the ground below. A huge explosion created a fireball rising to nearly 5,000 feet that was visible from more than 50 miles away. More explosions followed as firemen concentrated on cooling the railroad cars until the fire burned itself out. The greatest danger came from a ruptured tank car that had the potential for spreading a deadly cloud of chlorine gas over the city. Authorities quickly evacuated more than 200,000 people, the largest evacuation ever undertaken in North America due to a man-made disaster. The emergency response was successful, and no deaths resulted.

The Canadian mishap was an example of excellent response to a potential disaster. In other transportation spills as well, loss of life has been limited through the use of fast and effective evacuations, or by favorable weather conditions that quickly moved the toxic plume away from populated areas. But another Canadian town was not so fortunate:

In July 2013, a train on the US-based Montreal, Maine, and Atlantic Railway was carrying crude oil across Quebec. The oil had been obtained from fracking shale in North Dakota. The engineer parked the train for the night on a hillside seven miles from the village of Lac-Megantic so that he could get some sleep at a local hotel. Later that evening, a fire broke out in one of the locomotives, which the local fire department extinguished. In the confusing circumstances that followed, the train's 72 tank cars became uncoupled from the locomotives, and an insufficient number of the tank cars' hand brakes had been engaged.

Later that night, the brakes gave way completely, and the ghost train hurtled down the hill, reaching speeds of more than 60 miles per hour as it careened into Lac-Megantic seven miles away. The train derailed, massive amounts of oil spilled, and five or more of the cars exploded, creating a wall of fire reaching high into the sky. The blast decimated the town's center, destroying 30 buildings and forcing 2,000 residents to evacuate. The ensuing blaze was so intense and long-lasting that the bodies of many of the 47 who died could not be recovered until days later.

* * * * *

A highway disaster in Spain 35 years earlier was even more deadly than the Canadian rail conflagration. On July 11, 1978, a tanker truck left the state-owned Enpetrol refinery overloaded with 23 tons of highly flammable liquefied propylene and proceeded along a winding highway that passed through densely populated areas near Tarragona on its way to Barcelona. To avoid paying tolls, company officials instructed the driver not to take the more direct—and safer—expressway.

While the cause of the accident remains in dispute, for some reason the tanker developed a leak, causing the driver to stop next to the Los Alfaques campsite, which was crowded with nearly 1000 tourists in trailers and tents. A white cloud of gaseous propylene coming from the truck drifted over the campsite and on to a nearby discotheque. Campers, unaware of the danger, approached the cloud with curiosity. As the cloud reached the discotheque, it ignited, flashed back to rupture the tank truck, and caused a blast and fireball so powerful that it destroyed the campsite and discotheque, leaving 217 dead and severely burning 200 others. With no warning or time for evacuation, it was among the deadliest accidents attributable to technology to take place at that time.

* * * * *

The loss of life in the disasters detailed above is frightful. More fearsome yet, as technology advances, possible scenarios for even greater loss of life can be imagined. As national economies require the use of larger quantities of flammable, explosive and toxic materials, refinery and chemical processing plant inventories increase, oil and liquid natural gas tanker capacities become larger, and demand for pipeline, rail and highway transport of dangerous materials grows. As larger numbers of people concentrate in confined places, as seating capacities on everything from airliners and cruise ships to theaters, sports arenas and other structures increase, so too does the potential for disasters resulting in greater death tolls increase. Under slightly different circumstances, a number of the disasters described above could have led to even larger loss of life. Two near-misses in the US further demonstrate the point.

In 1979, a major storm with heavy rains and 70 mph winds blew through Kansas City, Missouri, and caused a partial roof collapse of the Kemper Arena. Fortunately, although the collapse took place at six forty-five in the evening, the indoor arena was not in use. Otherwise, with a seating capacity of 19,500, a large loss of life could have resulted. An even closer call was the Hartford Civic Center arena, which collapsed on January 18, 1978, under the weight of an unusually heavy snowfall. The collapse took place after the 16,300 capacity crowd had left after a college basketball game held only hours before. As such venues become larger, safety requires that building codes be strengthened, that warning signs of approaching danger be broadcast, and that effective evacuation procedures be in place.

As the size of ships—particularly cruise ships—grows larger, it is more and more urgent to examine more stringent safety precautions in the context of the larger potential loss of life. The fate of the Costa Concordia on the night of January 13, 2012, offers a cautionary tale. The captain's recklessness caused the ship to hit a reef off the Tuscany Coast of Italy and run aground on the Isola del Giglio. More than 30 people died in the tragedy as the ship came to rest partially submerged on its side. Had the vessel veered into deeper water and sunk rapidly instead of coming to rest in shallow water, the death toll would have been much larger, for there were more than 4,000 passengers and crew aboard the vessel.

Cruise ships currently being designed and launched are even larger, carrying passengers and crew of nearly 9,000! Recent events in which large cruise ships have experienced engine failures, fires, and other difficulties call attention to the need for improving methods for dealing with emergencies and for rescuing thousands of people should such a behemoth suddenly need evacuation.

* * * * *

The foregoing examples illustrate the devastating impact that disasters often have and bring to mind yet more catastrophic situations that could result from technological failures. Disasters attract intensive media coverage and imbed their horrific consequences in our collective memory. Investigations inevitably follow catastrophes, and investigative bodies make recommendations to reduce the risk that the tragedy will be repeated.

Putting in place the lesson learned, however, inevitably poses thorny issues: Is it sufficient to require that a new technology include the recommended safeguards, or do current equipment or structures have to be retrofitted? If retrofitting is required, how much time should be allowed for the work to be completed, and should the technology continue to be used while the work is completed? Finally, if it proves to be uneconomical or impossible to make adequate upgrades, should the buildings be torn down, the aircraft junked or the ships scuttled? The economic costs of the corrective measures must be weighed against the benefits gained as society strives to reduce technological risks.

Progress toward increased safety follows as lessons learned from tragic experiences of the past are incorporated into new technological systems. As time goes by, however, the construction of larger cruise ships, buildings with more and more residents, and stadiums with ever-increasing seating capacity concentrate more people in one location. The increased density of humanity in such venues multiplies the number of lives that could be lost should the worst possible disaster take place, whether it be a ship sinking, a building fire, or a stadium collapse.

To place the dangers of potential disasters into perspective, probability must be considered. Probability, however, has been a problem in risk perception since the word was first introduced in 1662 by a group of Parisian monks. They were puzzled over their contemporaries' fear of lightning even though the risk of being hit was very small compared to the other risks that they faced daily in that preindustrial society. It prompted the statement:

> Fear of harm ought to be proportional not merely to the gravity of the harm, but also to the probability of the event.

The problem of understanding probability persists today, coloring our perceptions of risk. We discount the risks stemming from frequent roadway accidents, falls, fire, and drownings that claim one or two victims at a time, and amplify fear of rare but disastrous accidents, even though in any given year, disasters take many fewer lives. In examining disasters, the probability of the events must be considered in tandem with the expected consequences.

Probabilities are abstract and difficult to comprehend, particularly when they are small. Our intuition often leads us astray, causing us to grossly overestimate the likelihood of some dreadful possibilities and to underestimate—or totally ignore—the chance that others will occur. We may have once joked about the rarity of being "struck by lightning." But now that lightening strike statistics have received more attention, we understand that lightning causes between 20 and 40 deaths in the US each year. Commercial airliner crashes, which many people fear much more, in some years cause fewer deaths than those attributed to lightning.

Looking at the number of lives lost in disastrous accidents provides some perspective in understanding the risk of catastrophic technological failures. Figure 10.1 displays the numbers of lives lost in man-made disasters in the United States since accident data collection started in the mid-nineteenth century.

For this analysis, a disaster is defined as an event caused by human activity with a death toll of more than ten lives. Not included are

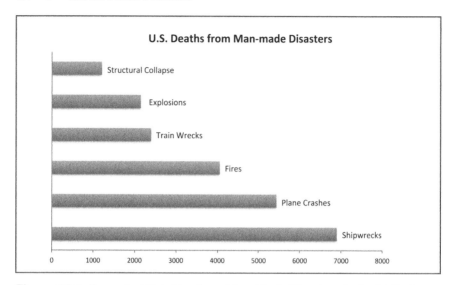

Figure 10.1: Recorded Fatalities from Man-Made Disasters in the United States, *Wikipedia*

natural disasters, whose death tolls tend to be larger and have sometimes been increased by failures of dams, levees, or other technology; these are examined in Chapter 11.

Observe that since records have become available, shipwrecks have taken the largest toll, followed successively by plane crashes, fires, train wrecks, explosions, mining accidents, and structural collapses. Only shipwrecks have resulted in catastrophes in which 1000 or more lives were taken in a single disaster. In fact, three maritime disasters—the *Titanic* the most famous of them—account for two thirds of the lives lost aboard ships in peacetime, and the most recent of these took place more than 100 years ago. Indeed, the risk of death from man-made disasters has greatly diminished over time. Two thirds of all the recorded disaster fatalities took place before 1950; if air travel is excluded, only about fifteen percent of the fatalities arise from disasters since 1950.

Perhaps the largest surprise to many comes with the comparison of disaster deaths to those from ordinary accidents. The total of all disaster deaths recorded in US history is roughly 24,000. This number is thousands less than the 35,000 lives lost *each year* in motor vehicle accidents. The per capita numbers are much higher for both disasters and motor vehicle accidents in developing countries where poverty is great and concern for safety minimal. But disasters also take fewer lives there than do the more mundane accidents that take lives one or two at a time.

Disasters elicit a level of dread far out of proportion to the number of lives lost. This becomes evident if we compare air and automobile travel. The deaths from commercial airline crashes in the US, most often far fewer than a hundred in a year, hardly compares to the annual rate of tens of thousands of lives lost in automobile accidents. The chance of perishing in a motor-vehicle accident is more than twenty times greater than traveling the same distance by scheduled airliner. And yet many who drive with abandon, and often without seat belts, are deathly afraid of commercial aviation. Why is there such a discrepancy in the public's risk aversion to these two classes of accidents? Why does one aircraft crash killing 100 people garner so much more attention than 100 automobile accidents each killing one person?

The spectacular nature of airliner crashes—the exploding fireball, wreckage strewn over miles, and the grim spectacle of collecting and identifying burned or dismembered human remains—brings forth greater horror than of police pulling one or two bodies from a wrecked automobile. News coverage is national, even international, with many frightful pictures, and coverage continues as the public waits for the National Transportation Safety Board to perform its investigation. Did equipment failure, pilot error, or perhaps an act of terrorism bring the plane down? The Safety Board methodically pieces together the evidence to determine probable cause and only then makes recommendations to improve aircraft design, pilot training, air traffic control, and other changes to lessen the chance of a reoccurrence. After that final report, the accident only gradually fades from the public consciousness.

The degree of control that motorists feel in driving their automobile compared to the lack of control felt by passengers on an airplane may be an even greater factor. Even though motorists may take to the roads with autos in poor repair, in bad weather, or when they are sleepy, distracted or drunk, they feel that they are in control. And should they begin to feel unsafe, they feel they can always slow down, drive more cautiously, or pull off the road. If anyone is fearful, it is more likely to be the passengers than the driver, but even they feel they can demand that the driver slow down or stop.

In contrast, airplane passengers are not in control. They may feel that they are stuffed inside a tube, tens of thousands of feet about the ground, traveling at nearly the speed of sound, where the air it too thin to breathe. They have no input as to how the captain flies the airplane or control over the weather conditions that the flight may encounter. To them it may be of no consequence that the aircraft has at least two engines and all manner of redundancy and back-up systems not found in automobiles to allow their flight to be completed safely even if multiple equipment failures do occur.

It matters little to them that, unlike the motorist, the captain has a copilot for backup, and that the two of them have had years of experience before taking charge of an aircraft. Nor do they think about the regular physical examinations the pilots must have, or the frequent drills in simulators they must perform on how to respond to engine failures, on-board fires, malfunctioning landing gear, adverse weather conditions, and a host of other unlikely emergencies. The lack of personal control is an overwhelming emotion that causes many to fear flying.

Psychologists and economists have expended a great deal of effort in attempting to understand how we form our perception of the risk that disasters present and how such perception relates to the realities of lives lost and physical destruction. Their studies indicate that we perceive the likelihood of an event by how easily a similar event comes to mind. Visions of deadly destruction imprinted indelibly in our memories by vivid media coverage heighten our feelings that similar events are likely to occur in the near future. The impact of disasters on our psyche greatly outweighs the calming effect of statistical studies that take into account the rarity of such events. Such studies inevitably conclude that the risk to our welfare from disasters is small compared to the more ordinary accidents that occur every day, and to the dangers to our health and safety that are much more within our control, but about which we think little.

* * * * *

Our dread of disasters no doubt exaggerates their importance in the risks that we fear relative to the mundane accidents on the highway, at home, and at work that are far more likely to kill or disable us. It seems likely that expending more resources to reduce the numbers of everyday accidents would do more to reduce accidental death tolls than would allocating comparable amounts of money to further reduce the probability and consequences of disasters.

Nevertheless, wise public policy requires that the avoidance of disasters receive continued scrutiny. Roughly speaking, the risk presented by a disaster may be measured as its probability multiplied by the number of lives lost. Thus, as conceivable death tolls from worst-possible disasters grow, efforts to reduce them must be aimed both at reducing the likelihood of catastrophic accidents, and at implementing policies for emergency preparedness that will decrease the consequences of disaster should they occur. Unlike the man-made disasters discussed thus far, humankind is powerless to reduce the frequency of hurricanes, earthquakes and other natural disasters. Conversely, technology is central to mitigating or exacerbating their consequences, as I shall discuss in the following chapter.

CHAPTER 11

Natural Disasters—Technology's Impacts

GALVESTON WAS A boomtown at the turn of the twentieth century, with a population of 37,000. Connected to the mainland by bridges, the city was located on a barrier island along the coast of Texas, 15 miles long and 3.5 miles across at its widest. In September 1900, wireless telegraphy was in its infancy, so the only warnings of a storm that was brewing in the Gulf of Mexico were relayed by seamen coming ashore. Moreover, at that time, the National Weather Bureau avoided use of the term hurricane for fear of panicking those in its path. Because the city had experienced at least eleven hurricanes in the nineteenth century, with the most severe having a storm surge of 8.2 feet, Galveston's citizens remained somewhat complacent.

On September 8, the storm reached Galveston, and it was record-breaking. Winds of 145 miles per hour pounded the city, and a storm surge of more than 15 feet swept over the island, drowning the city, which had a maximum elevation of less than nine feet above sea level.

The devastation was horrific; the hurricane destroyed nearly 4,000 homes, knocking them off their foundations and grinding them to rubble. The human toll was worse. An estimated 8,000 died, and 30,000 were left homeless, ranking the storm as the worst natural disaster recorded in US history. Most died immediately, drowned or crushed as water pulverized their homes. Others succumbed during the days that followed, trapped under building debris with rescuers unable to reach them.

Telegraph lines were down, bridges to the mainland destroyed, ships sunk, and railroad tracks washed away, making rescue attempts— or even getting word of the disaster's magnitude to the mainland— nearly impossible. On the morning following the surge, a captain and six passengers managed to cross Galveston Bay to the mainland on one of the few surviving ships. They made their way to the Houston telegraph office and sent a short message to the Texas governor and to US President McKinley, announcing that Galveston was in ruins.

The aftermath of the hurricane was gruesome. Bodies were decaying everywhere, and they were so numerous that they made prompt burial impossible. An attempt was made to weigh the bodies down and bury them at sea, but currents in the Gulf of Mexico washed too many of them ashore. The storm's survivors resorted to an alternative solution. They set up funeral pyres for cremation wherever bodies were found; the pyres continued to burn for weeks following the storm.

As Galveston recovered from the disaster, authorities determined that advances in building technology would permit a seawall to be built to protect the city. Construction began on the first section of the 17-foot Galveston Seawall in 1902. A strong hurricane tested the wall in 1915, with a storm surge three inches higher than the surge of 1900. For the most part, the wall succeeded in protecting the city; there was some flooding, but only eight Galveston residents lost their lives. Later improvements provided additional protection. The elevation of the city was raised, in places by as much at 16 feet, to further reduce flooding from future hurricanes.

Since the Galveston calamity more than a century ago, numerous natural disasters have struck within the borders of the United States, but none have been so deadly as the 1900 flood. Much worse natural disasters have befallen less fortunate countries; the more than three million lives claimed by the 1931 flood in China is the largest number of fatalities recorded in modern times.

* * * * *

Figure 11.1 displays the death tolls from the most deadly natural disasters (i.e., those taking more than 10 lives) recorded in the

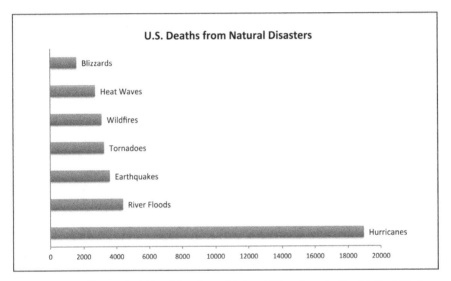

Figure 11.1: Recorded Fatalities from Natural Disasters in the United States, *Wikipedia*

United States since the mid-nineteenth century. Hurricanes are by far the largest contributors, followed by river flooding, earthquakes, tornadoes, wildfires, heat waves, and blizzards. All are weather-related except earthquakes, which will be considered later. Observe that the total death toll of about 38,000 for natural disasters is substantially higher than the 24,000 total for man-made disasters displayed in Figure 10.1, but roughly the same as the 35,000 motor vehicle deaths *each year*. Thus the death tolls from both made-made and natural disasters do not reach those of commonplace highway carnage.

With the exception of Hurricane Katrina in 2005, the most devastating US hurricanes (those taking 1000 or more lives) pounded coastlines in the nineteenth and early twentieth centuries. Advancing technology has played an important role in reducing the deaths caused by hurricanes. Hurricane warning systems, aided by improving satellite prediction of storm paths and rapid communications, allow people to begin evacuating low-lying areas much earlier and have lessened the numbers of deaths. Moreover, evacuation can be accomplished more rapidly because motor vehicles and paved highways have replaced horses traveling over dirt paths; fast boats, helicopters, and other motorized rescue equipment now aid in rescue efforts.

Present-day communications are capable of giving warning for nearly all weather-related natural disasters, ranging from days for hurricanes to minutes for tornados. But technology alone is insufficient. Public officials must have sheltering and evacuation plans in place well in advance, and the public must be educated to abide by them. Evacuation efforts are thwarted if unreliable predictions result in too many false alarms, resulting in the public tending to ignore the alerts. And if a catastrophic event has not occurred in recent memory, complacency becomes an increasing problem.

Even with increased numbers of citizens choosing to live in locations at high risk for hurricane damage, technological advances have dramatically reduced the death tolls from hurricanes. In the last half century, only four hurricanes have caused more than 100 deaths, while the others have caused substantially fewer. Hurricane Katrina was the exception, and to a large extent the death toll of 1,600 can be blamed on the technological breakdowns detailed in Chapter 6, most importantly the catastrophic failures of the levee system intended to protect New Orleans and the faulty evacuation plans.

The death tolls from river flooding, second only to hurricanes in causing weather-related deaths, have been reduced over recent decades as a result of many improvements in warning and evacuation technology. A striking feature of the flood data, however, is that by far the most deadly flood resulted from a failure of a nineteenth century earthen dam. Excessive rain in the spring of 1889 caused the reservoir above the South Fork Dam in western Pennsylvania to overfill. On the morning of May 31, the dam eroded and collapsed, sending a torrent of water, which at its peak equaled the flow of the Mississippi, down the narrow Conemaugh River valley. Johnstown, a city of 30,000, was located fourteen miles downstream from the dam. The wall of water hit without warning, inundating the city and causing 2,200 deaths, ranking it among the worst disasters in US history.

Water control using levees, sea walls, and dams has played a substantial role in keeping flooding from populated areas and saving lives. As hurricane Katrina and the Johnstown flood illustrate, however, such structures have their drawbacks: if flooding is greater than they can handle, water may overflow them, or worse, the structure may be breached or collapse, causing even greater loss of life than had it not been built. Wherever a dam, levee, or seawall is built, serious consideration should be given not only to the probability that it may be overwhelmed or destroyed by a storm of unprecedented strength, but also of what the warning signs would be of such an approaching storm, and what preparation is needed to carry out a successful evacuation.

Improved construction technology has facilitated making homes more resistant to wind damage from hurricanes and building shelters from tornadoes. Situations differ depending on the threat. Protection from deep flooding is difficult. Even though technical measures may reduce wind damage, evacuation is often the only solution when hurricanes inundate low-lying areas. Likewise, only evacuation may save lives from widespread flooding of the Mississippi or other major rivers. Conversely, there is rarely time for evacuation when a tornado alert is sounded. Shelters must be in place and must be accessible in minutes, if not seconds.

In building safe havens, public officials and homeowners face questions of costs and benefits. For areas of the country where tornadoes are frequent, homeowners must ask themselves if the chance that their particular houses will be flattened by a tornado is large enough to justify the expense of constructing storm shelters. Likewise, public officials must decide whether and how to put shelter requirements into building codes.

* * * * *

Unlike weather-related natural disasters, earthquakes give no forewarning. Such was the quake that rattled San Francisco at five in the morning on July 18, 1906. The shaking lasted less than a minute, but it toppled chimneys, collapsed walls, buckled cable car tracks, and ruptured water mains and gas pipelines. But worse was yet to follow. Fed by the predominance of wood construction and the ruptured pipes that supplied the city's extensive gas lighting, fires erupted immediately following the quake and spread to consume 90 percent of the city. With the water mains destroyed, the fire department could do little.

Army units were called in from nearby outposts, and the troops dynamited blocks of buildings hoping to create a fire break, but the tactic failed to thwart the spreading inferno. The military's presence did prevent looting and made evacuation more orderly as waves of the city's inhabitants—with what belongings they could carry—fled the city on foot ahead of the advancing walls of flames. Three days elapsed before rain came and the fire burned out.

The San Francisco earthquake and fire caused 3,000 deaths, destroyed nearly 30,000 buildings, and left more than 200,000 homeless. Tent cities were set up, and surrounding communities accommodated many of the refugees. No one went hungry, and outbreaks of disease were avoided. The city's resilience displayed itself in the rapid rebuilding that followed the disaster, and the science of seismic design was born in the aftermath of the earthquake as engineers studied the damage suffered by the city's buildings and infrastructure.

Business and political leaders, however, prevented references to earthquakes in the revised building codes. In a publicity campaign, they blamed the destruction on the fire, fearing that East Coast investors would be frightened away by the fear of future earthquakes. Major advances in incorporating seismic design criteria into building codes were delayed until the 1925 Santa Barbara earthquake caused widespread structural damage. Since then, the development of earthquake warning systems and increased knowledge of how to construct buildings, bridges and other infrastructure to minimize seismic damage have greatly reduced human casualties from more recent earthquakes.

In seismic terminology, a warning must be distinguished from a prediction. There is no way to predict when or where an earthquake will occur, nor is there likely to be a solution to this problem in the foreseeable future. An earthquake warning is something different; it is an alert that an earthquake has begun. The alert arrives at a location some distance from the quake's epicenter before the damaging earthshaking begins.

Earthquakes radiate p waves and s waves though the earth. p waves travel faster than s waves, therefore arriving before the s waves. But it is the slower-traveling s waves that cause destructive ground shaking. Thus, if the arrival of the first p wave is detected, the time interval between the arrival of the p and s waves at any location serves as a warning. The time interval between p and s wave arrival varies between several seconds and a few minutes, depending to the strength of the earthquake and the distance from its epicenter where the warning is activated.

Even with forewarnings of only seconds or minutes, much can be done: high-speed trains can be slowed or stopped, valves on gas and oil lines closed, landing aircraft diverted, elevators emptied, and other potentially dangerous machinery shut down. Within several seconds of the sounding of school alarms, children can reach protection under tables; when earthquake warnings are spread using cell phone apps, much of the public could use the seconds or minutes available to seek cover. Japan and Mexico are leaders in installing such systems, which are also being installed in seismically active parts of the US and other countries, though at a slower rate.

Other technological advances over past decades have served to reduce earthquake death tolls. The sturdiness of buildings has improved, along with fire protection and rescue techniques. But the sizes of our cities have grown, and as other more intense quakes in other countries remind us, a more devastating quake is likely—indeed is certain—to befall the US.

Builders of buildings and bridges, power stations, waste disposal facilities and many other structures must take the eventuality of devastating earthquakes into account. And this is yet another context for determining how safe is safe enough. Extensive research has provided a better understanding of earthquakes. Scientists can make predictions of the energy generated—measured on the familiar Richter scale—as well as of the frequency at which quakes of a given magnitude are likely to occur in a particular region. Extensive study allows seismologists to estimate the strength of the quake likely to take place once in a decade or once in a century, where the longer interval corresponds to the more devastating quake. Such predictions then make possible estimates of the intensity of ground shaking that will likely be associated with each location, for it is ground shaking that causes structural damage or collapse.

But what level of ground shaking should engineers design structures to withstand? The problem again is in dealing with probabilities. The intensity of an earthquake expected once in fifty years is stronger than one expected once in ten, and a quake expected once in a century is stronger yet, and so on. The question is again how safe is safe enough, and some form of cost benefit analysis must underlie the decisions. Common buildings may be designed to withstand a once-in-500-year earthquake; auditoriums, monuments, dams, waste disposal sites, bridges, and other structures for which severe damage could lead to greater loss of life should be designed to withstand even more severe earthquakes. But in no case can it be said that they are completely immune from earthquake destruction.

Authorities must determine whether the additional number of lives that could be saved can justify the added cost of reinforcing a structure to withstand a more powerful earthquake. But this requires, once again, that a monetary value be placed on a life. Such decisions are by no means the prerogative of those who design and build. The public's input comes through government agencies that stipulate what measures must be taken. In this way, the political process works its way through to create the building codes that governmental bodies mandate.

It is not only in the design of new structures where earthquake risk must be considered. Retrofitting is also a large issue for schools, stadiums, multi-unit housing, and other buildings that are not replaced frequently. Currently, 50 years is a typical life for which such buildings tend to be designed. But many buildings much older than that remain in use. As earthquakes have become better understood, the level of protection demanded has increased markedly. If a fixed budget is available, and a governmental body is trying to decide which project to undertake, the question then becomes not what is the value of a life, but simply which project will save the most lives for the money available.

Inevitably, determining acceptable levels of earthquake protection is a contentious process. In Los Angles, for example, plans to build two skyscrapers have been delayed by lawsuits arguing that the structures may straddle an earthquake fault line. Retrofitting has also become a disputed issue, triggered by a University of California study indicating the vulnerability of 1,500 concrete buildings in Los Angeles built before 1975, when a stricter building code went into effect. Just as important, many of the buildings were built before a number of nearby fault lines had been discovered.

Retrofitting—and how to pay for it—has set seismic safety against financial interest, and geologists against business owners. Indeed, those responsible for the University of California report feared that its release would create excessive alarm and cause a precipitous drop in property values. In many cases, the $100,000 or more required just to do a thorough inspection of a building might exceed its market value. Retrofitting requirements have been strongly resisted by owners of the older buildings, often setting their financial interests against the safety concerns of those who actually live in the structures. Such are the practicalities in coming to decisions on how much protection against rare but devastating earthquakes can be justified and who should pay for it, given the large financial cost of implementation.

Economics places limits on how improbable an earthquake, flood or other natural disaster a society can require its technology to withstand. Is a natural disaster estimated to occur on average once in five hundred or once in a thousand years too remote and unaffordable to consider? Such an event may be exceedingly rare, but it is not impossible. And should it occur, causing structures to be damaged, collapsed or destroyed, are there ways to limit the loss of life and other dire effects that may otherwise follow? The failure of the New Orleans levee system that resulted from hurricane Katrina forced officials at all levels of government to confront these questions. Likewise such questions are currently being confronted in Japan following the devastating 2011 earthquake and tsunami.

* * * * *

On March 11, 2011, many of the issues concerning the impacts of technology on natural disasters came together. An earthquake of 9.0 on the Richter scale broke out 80 miles off the eastern coast of northern Japan. It was the largest recorded in the country's history. Strict building codes and other safeguards protected the population quite well from the severe ground shaking caused by the quake. However, a tsunami of epic proportions ensued, reached the coast 55 minutes later, and destroyed the sea wall that had been built to protect Fukushima Prefecture. The wave brought unimaginable death and destruction, washing away buildings, drowning inhabitants, and leaving nearly 20,000 dead or missing!

Arguably, the sea wall—which had offered protection from lesser tsunamis—added to the death toll, for the wall's presence gave residents of the coastal area a false sense of security. If the authorities had relied less on the wall for protection, they might have required stronger structures built at higher elevations to serve as nearby safe havens, instituted more rapid escape routes, and limited development in the lowest-lying areas.

At Fukushima, the protective seawall was only 19 feet high. But the tsunami reached heights of over 40 feet, overwashing and destroying it. Authorities had deemed the wall to be high enough to protect the city and the nearby nuclear power plant from any foreseeable tsunami.

In the 1960s, when the reactors were built, no tsunami that high had appeared in the historical records that they examined. A more thorough study in 2008, which took a 1896 earthquake and tsunami in the region into account, indicated that tsunamis as high at 34 feet were a possibility, and pointed out the immediate need to improve protection. The power company, however, resisted the high cost of protection against such a wave, and the government did not force the issue.

The six nuclear reactors at Fukushima-Diachi were very different from the Soviet-built plant that exploded at Chernobyl. Built in the 1970s, they conformed to standards comparable to those required at that time in the United States and Western Europe. But those other plants were not located where tsunamis were a threat. The Fukushima reactors' reinforced structures protected them from the severe ground shaking that inevitably accompanies the earthquakes to which Japan is subjected. Each had diverse and redundant safety systems to assure that, in any emergency, the energy-creating nuclear fission chain reaction would be shut down immediately.

The designs provided multiple paths for the safe removal of the so-called decay heat generated during the days and months following shutdown by the radioactivity of the fission products encased in the reactor fuel. In addition, each reactor was contained within multiple containment barriers to prevent the radioactive fission products from reaching the environment, even in an emergency. Each reactor core was located inside a steel pressure vessel through which water circulated to provide cooling. The vessel was located within a larger steel structure—called the primary containment—that was also capable of withstanding high pressure. This primary containment was located within a larger building—referred to as the outer containment—that served as a third and final barrier between the reactor's large amounts of radioactive materials and the outside world.

The Achilles heel of the Fukushima facility was a combination of the inadequate protection provided by the sea wall to shield the reactors from tsunamis and the placement of emergency equipment in basements in a flood-prone area. Without the tsunami, what damage the reactors incurred from the earthquake's ground shaking would have been manageable, and radioactivity would have been contained within the reactors' cores. But when the seawall failed, the devastating wall of water flooded the reactor buildings because the company had considered the possibility of such a tsunami to be too remote to take adequate preventative precautions against flooding.

Since the reactors had been shut down before the tsunami overpowered the seawall, the challenge was to remove the decay heat over the long term. Otherwise, the reactors would eventually melt, and if the containment barriers failed, large amounts of radioactive materials could escape to the environment. But the tsunami knocked the region's power lines down, and electricity was required to operate the pumps, valves, and other equipment needed for orderly removal of the decay heat.

Contingency plans took power failures into account, calling for battery power to suffice for a short time until redundant sets of emergency diesel generators would start up and take over. But the flooding put the generators and associated fuel tanks—located in basements—underwater and incapable of functioning. The resulting loss of cooling capability set off a disastrous sequence of events.

As the realization became widespread that the reactor cores would eventually melt without emergency cooling, media worldwide reported the ongoing drama of the stricken reactors nonstop. Attention rapidly shifted from the 20,000 victims drowned or missing in the immediate aftermath of the tsunami to the ominous developments at the nuclear power plant. Like the aftermath of a plane crash, train wreck, or fire, the danger of immediate death was over once the tsunami had passed. The media quickly turned its attention to the slowly unfolding drama of the state of the crippled reactors and the foreboding of radioactive release and its ominous health consequences. It was a compelling story indeed.

As the seriousness of the situation became apparent, government officials ordered citizens living within two miles of the reactor to evacuate. Meanwhile, arrival of replacement batteries and portable generators for the reactors was greatly delayed due to the severe destruction that the tsunami caused to the entire coastal region. Even after emergency equipment arrived, attempts to provide emergency power failed because the flooding had also disabled vital connecting equipment and made the equipment impossible to access for repair.

On the morning following the quake, authorities ordered an evacuation of those living within six miles of the reactors. At the same time, efforts to prevent reactor meltdowns shifted to stringing temporary lines from the disabled power grid to the reactor coolant pumps. Successful connection to that grid only took place much later, however, so late that it was of little value in decreasing the most destructive phases of the accident.

Without resupply, the water in the pressure vessels boiled off, eventually uncovering three of the reactor cores. As the steam surrounding the tens of thousands of fuel rods in each reactor became hotter, it interacted with the zirconium metal tubes, which contained the radioactive fuel, and the chemical reaction generated hydrogen gas. The steam and hydrogen raised the pressure within the reactor vessels to excessive levels, causing safety valves to open, relieving the pressure, but allowing the hydrogen to escape from the reactor vessels into the primary containment.

The reactors were constructed to vent hydrogen through tall external stacks designed to filter or burn the hydrogen in a controlled manner before it entered the atmosphere. But the filtering systems didn't work without electricity. Instead, the hydrogen flowed into the secondary containment, where it mixed with air and ignited.

The hydrogen explosion blew open the secondary containment building of one of the reactors the day after the earthquake, and two days later hydrogen exploded in two of the plant's other reactors, with similar results. Captured on camera, the explosions were spectacular, throwing debris high into the air and creating mushroom-like clouds; the strongest was felt more than 30 miles away. The explosions heightened fear of the release of massive amounts of the contaminant most feared by the public—radioactive material.

The explosions allowed gaseous fission products that had been vented along with the hydrogen to gain direct access to the atmosphere. The spike of atmospheric radiation that ensued prompted the government to expand the mandated evacuation zone to the 150,000

Figure 11.2: Hydrogen Explosion at the Fukushima Nuclear Power Plant, *Associated Press*

people living within 12 miles of the reactors and recommend that those living between 12 and 18 miles from the plant remain indoors to minimize their exposure. As the uranium in the cores of three of the reactors melted, an ongoing stream of radioactive material escaped to the atmosphere. Two weeks later, the government recommended that 350,000 residents of this outer zone relocate further from the plant.

The hydrogen explosions' destruction of the secondary containment buildings created an additional threat. Those buildings contained spent fuel, older fuel rods that had been removed from the reactor and stored in deep pools of water. The water was necessary both to shield the environment from radiation emitted by the rods and to provide cooling for the decay heat generated by the longer-lived fission products. With the cooling system lost, the danger was that the heat would eventually cause the pools to boil dry, and the spent fuel to melt. Such melting would progress at a much slower rate than the melting of the reactors' cores since the stored fuel rods generated much less decay heat. But without water in the pools, the radioactivity would be directly exposed to the atmosphere.

Frantic efforts continued to cool both reactor cores and spent-fuel pools. On the day following the accident, a jerry-rigged system began injecting seawater into the primary containment of one of the reactors in the hope of slowing or stopping its temperature rise. Subsequently, seawater was injected into two of the other reactors as well. With instrumentation destroyed, operators had difficulty understanding what was happening inside the reactor vessels. They suspected fuel was melting in three of the reactors; following the hydrogen explosions, their fear of the spent-fuel pools boiling dry increased. Not until the sixth day into the accident were helicopters able to dump seawater into any of the spent fuel pools. Later, emergency crews succeeded in adapting riot control water cannons and specialized fire-fighting equipment to spray seawater from above onto the pools. More than a week elapsed before a temporary distribution system began to supply offsite electricity to the nuclear plant.

Three of the reactors at Fukushima released sizable amounts of radioactive material to the environment, peaking after the hydrogen explosions and diminishing slowly during the weeks following the tsunami. Yet no one, not even the reactor operators or emergency workers, received enough exposure to show any signs of radiation sickness. In fact, the destruction of the reactors caused only three deaths. Two operators drowned and one died from the collapse of a crane.

But that was no cause for relief among the Japanese, or even for many who lived oceans away. Imminent death wasn't the issue, but dread of how many would perish in the coming years from delayed health effects, the cancerous effects from the cloud of radioactive material that spread over the surrounding countryside, contaminating soil and sea, and carried far and wide in detectable amounts to other countries and other continents.

* * * * *

Radioactive material and the radiation emitted from it are among the most feared forms of toxic contamination. The dangers from nuclear radiation have been studied intensively and are well understood. But for nearly three quarters of a century, the mushroom cloud that hung over Hiroshima following the US dropping of the atomic bomb continues to hang over the public's collective psyche, heightening fear of radiation, often out of all proportion to the risks that it presents.

Thus around the world, the most dreaded consequence of the Fukushima disaster was not the tragedy of nearly 20,000 drowned or swept away by the force of the tsunami, but the radioactivity released from the tsunami-damaged nuclear plant. Had radioactive materials not been present, the attention given to the destruction of a large industrial plant would have been scant in the wake of a massive natural disaster of such historical proportions. What took place in the extended battle at Fukushima to stem the release of radioactivity to the environment was spellbinding. It captured the world's attention and exacerbated its fears.

Radioactivity contaminated the landscape, polluted the ocean near the reactors and spread as the gaseous plume dispersed in the atmosphere. The greatest share of the exposure to the local population came during the week following the accident, but the contamination from the plume's fallout left a longer-lasting residuum of radioactive contaminants on the ground.

Attention was riveted on the accident and remained so as the drama unfolded in the weeks that followed. News media called in experts to comment on the ongoing events, but they had difficulty in saying anything significant because no one knew exactly what was happening inside the plant. Media debates emerged as to what course of action should be followed.

Worry extended far beyond Fukushima and its surroundings. In America, it focused on whether the radioactive cloud might reach the west coast, and whether contaminated ocean water might make eating fish caught in the Pacific unsafe. Many as far away as the east coast of the US—nearly 7,000 miles away—wondered whether they should be taking potassium iodide pills to prevent radioactive iodine from collecting in their thyroid glands. Their fear caused a run on iodine pills at many pharmacies in the US.

In response to the public's reaction to the accident, the Japanese government ordered the shutdown of all of its nuclear reactors—reactors that supplied more than a quarter of the county's electricity. Likewise, many other countries ordered emergency inspections of their reactors, and some, such as Germany, passed legislation to phase out nuclear power generation entirely. The economic cost for such reactions was severe, and the setbacks they caused to efforts to stem global warming have been immense. In Japan, particularly, the economy suffered. Office workers in Tokyo sweltered through the high summer temperatures without air conditioning, and some industries had to close down, later threatening to move their manufacturing plants offshore if the government did not allow reactors in key locations to restart.

Public fear and the political and economic repercussions stemming from the damage done by the tsunami to the reactors at Fukushima indeed had massive global impacts. Yet, exactly one year after the accident, Gregory B. Jaczko, the Chairman of the US Nuclear Regulatory Commission, speaking before an international gathering of more than 3,000 experts, stated that the accident's immediate health consequences were "very close to nothing," with "little to no long-term health effects." How could this be? The world's reaction to an accident had not been as intense since the ones at Bhopal and Chernobyl decades earlier. Still, the highest official in the US government's nuclear regulatory body deemed the accident's heath effects to be inconsequential! The Fukushima accident, perhaps more than any other, illustrates the dichotomy that sometimes exists between the public's views of technological risk and those of professional risk assessors. To understand this dichotomy, we must examine radiation risks in more detail.

Contrary to widespread impression, no one—not even the workers inside the Fukushima reactor buildings—was exposed to enough radiation to become sick. Instead, the danger was in an increased cancer risk. Two classes of cancer risk determine the consequences of nuclear accidents.

The first is exposure to radiation from radioactive material in the plume emitted from the plant as well as from the fallout and contaminants on the ground. Such radiation exposure comes from radioactive materials located outside the human body. It is called a *whole-body dose* or *external dose* because the radiation penetrates the body nearly uniformly. The second risk comes from the radioactive material that is inhaled or ingested and tends to collect in particular organs, in bones or the thyroid for example. It is called *internal dose*. We will consider the whole-body dose first.

A comprehensive study performed by the World Health Organization (WHO) estimated the amount of radiation exposure to the 150,000 residents who lived in the most exposed localities if they had not been evacuated. WHO estimated the exposure of those living between 12 and 18 miles from the reactors and those living more than 18 miles away.

Converted to cancer risk, the exposure of those in the evacuation zone, had they not been evacuated, would have resulted in a probability of 0.0015 of contracting cancer, with much smaller probabilities for those evacuated or living further from the plant. WHO concluded that, without evacuation, an estimated 225 cancers would be expected in the Fukushima locality, but with evacuation, less than 100 cases.

To place the projected incidence of cancer caused by the accident in perspective, consider that an average of 20 percent of the population would contract cancer even if there had been no accident. Thus in a population of 150,000 living closest to the reactors, 30,000 cancers would be expected were there no exposure from the reactors. And *if they had not been evacuated*, that number would have increased from 30,000 to 30,225 cases, an increase of less than one percent.

One might argue that any increase is unacceptable. But cancer rates vary from place to place by as much as 15 percent above and below the average incidence. Thus the 225 increased cases should also be compared to the natural variation of 9000 cancers. Since very few people consider such variations when deciding where to live or work, it seems that the much smaller risks associated with the accident should not be of great concern.

A second comparison adds perspective to the cancer risk from the accident. We are all bathed in radiation. This natural background arises from naturally occurring radioactivity in rock, sand, soil, and cosmic rays coming from outer space. Thus background radiation varies with how high above sea level we live, and on the type of soil under our houses and places of work. For example, altitude and rock composition cause background radiation to be nearly twenty percent greater on the Colorado Plateau than in the New England states. Moreover, while the US dosage from medical procedures varies greatly, on average it is comparable to that received from natural background radiation.

But how do these levels of radiation affect the incidence of cancer? The scientific evidence indicates very little, if at all. The numbers of cancers caused by background radiation—or from those caused by the reactor accident—are far too small to be detected among the 20 percent of the population who contract cancer from all causes. Other statistics add to the evidence that low levels of radiation have inconsequential effects of cancer rates. For example, it turns out that the cancer rate is 15 percent lower in Colorado than in New England even though the background radiation is nearly 20 percent higher! This does not mean that radiation prevents cancer, but simply that the effects of genetics, diet, and other lifestyle factors appear to totally mask any minuscule effect that whole-body radiation comparable to background levels may have.

Radioactivity that is ingested and inhaled must also be carefully examined when evaluating the cancer risk from an internal dose of radiation. The primary internal dose from reactor accidents derives from the iodine settling on the ground, where plants take up some of it and are subsequently eaten by humans or farm animals. The iodine thus ingested by cows concentrates in their milk and presents the primary risk to rapidly growing children and teenagers. If they drink contaminated milk, the radioactive iodine concentrates in their thyroid gland, where it collects and may eventually cause cancer.

Fortunately, the iodine has a half-life of only 8 days, which means that only half of it remains after 8 days, one quarter after 16 days, one eighth after 24 days and so on. After a few months, the amounts remaining are negligible. Thus if authorities take proper precautions, the threat of thyroid cancer can be greatly diminished. Potassium iodide pills—if taken before the iodine is ingested—saturate the thyroid with nonradioactive iodine, so that when the radioactive iodine is ingested, it will not collect in the thyroid. But even without these pills, simply quarantining food products—particularly milk—from contaminated areas greatly decreases the risk of cancer.

The Chernobyl accident, described in Chapter 6, was by any measure a far worse accident. It released at least ten times as much radioactive material to the atmosphere as did the accident at Fukushima. There, where no precautions were taken, the only measurable public heath consequence of the radiation released to the public was a rash of thyroid cancer, predominately in children.

Experts attribute the thyroid cancer epidemic following Chernobyl accident largely to the government's mishandling of the accident in the waning days of the Soviet Union. The local population was not informed of the severity of the accident until much damage had been done, and children in that impoverished area continued to drink the radioactive milk, which was by far the largest contributor to thyroid cancer. Fortunately, thyroid cancer is very treatable; thus, fewer than 20 deaths have resulted.

In contrast, the Japanese authorities quickly evacuated localities near the reactors and banned the contaminated milk and other food products. As a result, medical screenings performed on children and adults to date indicate that radiation exposure to the thyroid gland is well below levels considered to be dangerous by the physicians.

* * * * *

Emergency planning, particularly with regard to sheltering, evacuation and decontamination, also can benefit from the lessons of Fukushima.

Controversy has arisen since the accident as to whether the evacuation saved lives or whether it may have increased the number of deaths resulting from the accident. First, it is important to observe that an alternative to evacuation is to leave residents in place, provided they remain indoors with the windows closed. That precaution would have reduced both the external and the inhaled radiation doses to well below what would otherwise have been received by the population without evacuation, and thus likely reduce the estimated number of cancers. It would also have decreased the psychological trauma and physical stress experienced by those undergoing rapid—and to some extent chaotic—evacuation in addition to the already harrowing experience of the tsunami.

More than 20 hospitals and nursing care facilities were located within the 12-mile evacuation zone, housing more than 2,000 patients and residents. Evacuating these facilities hurriedly without knowing where the occupants could be taken greatly aggravated stress among the aged. Ten or more died in the vehicles in which they were being transported, and later estimates were that more than 400 deaths, predominantly among the elderly, resulted from the stress of evacuation and conditions in the relocation centers—more deaths than the estimates for the number of radiation-induced cancers that would have occurred without evacuation.

Eventually, the contaminated areas around the nuclear power plant will need to be decontaminated and resettled. In fact, the largest effect of a catastrophic nuclear accident is likely not to be loss of life from radiation exposure, but the disruption, economic loss, and psychological stress coming from the separation of a large population from their homes, with only a distant hope of returning. Major efforts are underway at Fukushima to decontaminate the land to the point where residents can return and eventually gain some sense of normalcy.

In those areas not swept away by the tsunami, the houses remain standing and the infrastructure in place, but radiation levels remain too high for the occupants to return. The announced goal of the government is to decontaminate the evacuated zones to levels that are comparable to the level of background radiation. A risk assessor would likely ask whether the large economic cost of decontaminating to such low levels, along with the trauma to the population from continuing displacement, is worth reducing the risk of the very few radiation-caused cancers expected to occur sometime in the future.

The public deserves sympathy; as the accident developed, very little information was available on exactly what was going on or how great the danger was. Many—not only residents of Fukushima and the surrounding countryside, but everyone on Earth—got the impression that the risk of cancer had been increased significantly; for those who are diagnosed with cancer in the coming years, the question will remain whether the nuclear accident was responsible. At the same time, comprehensive analysis by WHO and other scientific bodies concludes that the likelihood of anyone not living near the reactors contracting cancer stemming from the radiation released is vanishingly small.

CHAPTER 12

Health Hazards—Foreboding Futures

As DEVASTATING AS shipwrecks, plane crashes, earthquakes, floods or fires may be, they share the characteristic that their death and destruction are apparent within a short period; those who survive need not fear future consequences. The situation is different for those exposed to toxic chemicals resulting from accident or routine use. The consequences may be poorly understood and often do not appear until much later in the form of cancer, respiratory failure, or neurological deterioration. Even contaminants—such as radiation—whose effects are well understood by experts, remain mysterious to the public, causing widespread fear. Cell phones present an analogous situation.

Mobile phones have revolutionized personal communications, bringing economic benefits, timesaving, and convenience to the hundreds of millions of people who use them worldwide. However, with such widespread use of hand-held phones, concern has increased that holding these powerful radio frequency (RF) transmitters against the head for long periods of time may lead to brain cancer or other adverse health effects. Thus the introduction of cell phones, like the introduction of synthetic chemicals and other technological advances, demands a concerted effort to assess possible adverse side effects early on; otherwise, such effects may go undetected for years.

Ferreting out adverse health effects can be a prodigious task, particularly if the effect is delayed or the occurrence is small. Already, well over a thousand scientific papers have reported research aimed at the potential health effects from RF radiation on the brain. The three paths of investigation parallel those employed to understand the health effects of ionizing radiation, pharmaceuticals, synthetic chemicals and other agents to which the public may be exposed.

Studies at the atomic, molecular and cellular level—the most basic levels—attempt to understand how the new technology may damage DNA, alter cell division or cause other physiological abnormalities. Laboratory studies using cell cultures and theoretical investigations, for example, have greatly deepened the understanding of how X-rays and other higher-energy ionizing radiation damage cells and increase the risk of cancer. Likewise, molecular biologists are learning how various classes of chemical compounds interact with living tissue. The considerable effort expended to date, however, has not shown how RF radiation exposure might cause cancer.

Epidemiological studies examine those who have been exposed to the suspected carcinogen to determine whether they exhibit a statistically significant increase in cancer or other illnesses compared to the general population. Studies of smaller groups that have had very high exposure, most frequently stemming from their jobs or from accidents, or whose exposure was long ago, giving cancer time to develop, are particularly valuable. Numerous studies have centered on health effects in research workers who worked on radar development during World War II, on naval personnel routinely working with high-powered RF emitters during the Korean War, and on amateur radio operators.

Studies have also been conducted on much larger numbers of cell phone users and on brain cancer patients to determine if they included an unusual number of major cell phone users. Such studies are made more difficult by the normal incidence of cancers, which vary significantly between economic, ethnic, environmental and other groups of human populations. Unless those exposed to RF radiation are carefully matched against a group of nonusers who are very similar in all these respects, differences in cancer occurrence may be due to other factors that are difficult to distinguish unless the incidence from RF exposure is quite large.

Animal studies using mice, monkeys, or other subjects may be conducted under much tighter controls, eliminating many of the factors that complicate epidemiological studies. Investigators take all of the animals from the same genetic line and raise them in identical environments so that there are no external causes for differences in cancer rates between the exposed and the unexposed control group. Information can be gathered more rapidly by continuously exposing animals to very high-intensity RF radiation 24 hours per day over a specific period. But if an effect is found, complications arise in how to extend the findings to humans, particularly if the animals have been exposed to many times more radiation than that of cell phone users. On the other hand, if no effect is found in the animals, can the same be said of humans?

In both epidemiological and animal studies, the smaller the effect—that is, the less likely cancer is to be induced—the larger the population of human subjects or animal specimens required to detect a statistically significant result above the background of normally occurring cancers. If an animal's chance of developing a tumor is one in a thousand, for example, then tens of thousands of animals must be exposed. Otherwise, the resulting cancer could not be distinguished statistically from one occurring purely by chance.

Ideally, investigators first try to determine whether a risk exists. If it does, the next step is to estimate the magnitude of the effect: What is the probability that someone using a cell phone will contract cancer? When this is known, the safety of the new product can be estimated, and then debate will ensue as to whether the product should be banned or how strongly it should be regulated to assure that the risk is acceptably small. With very rare exceptions, the thousands of studies related to RF radiation have found no evidence of an increased incidence of cancer. Nevertheless, investigators cannot say that there is no risk, only that if there is, it must be very small. To demonstrate that the risk is even smaller, they would need to conduct epidemiological investigations or animal studies with even larger numbers of subjects; these investigations would cost even more money.

In epidemiological and animal studies alike, investigators look for statistical effects that may be contradicted by other factors. Such studies are difficult to interpret. Many studies have been carried out looking for health effects from RF radiation; in virtually all of them, no effects have been found. Nevertheless, if even one study finds a seemingly significant effect, it will receive a great deal of publicity, and its principal investigator can plan on intense media attention. It matters little that the study may turn out to be a false positive, a statistical aberration, or some other fault that seems to indicate that there is an effect when there is none. The ongoing media attention will amplify public anxiety and create pressure for regulation, even though the broad consensus of scientific opinion may be that there is no cause for concern.

Cell phone use can be modified to subject the users to less RF radiation, using transmitters that are not hand-held to the ear but are at the end of a cord. If government agencies or scientific bodies recommend such precautions, however, the public would likely take the precaution as evidence of a larger danger, when in fact none had been found. These are difficulties that go beyond science. They become public policy issues in dealing with new technologies, particularly those that some think may lead to long-term health effects.

The scientific question of how safe the technology is becomes closely entwined with the public policy issue of what is safe enough. If public demand is to make it safer, then what level of regulation will reduce the risk sufficiently without creating economic impacts out of proportion to the small, and even questionable, gains in public health? All such issues come together in the creation of new technology. For example, how powerful can the RF transmitter in a cell phone be? The design rule must balance the potential risk from higher power with the longer range and greater distance between base towers that the more powerful handsets would allow.

* * * * *

Does it really make sense to expend resources to reduce what is a minuscule or nonexistent risk of RF-induced disease while accepting the far greater risk of talking on a cell phone—or worse, of texting—while driving? The National Highway Traffic Safety Administration estimates that drivers' use of cell phones contributes to 3,000 traffic deaths per year and that the delay in driver reaction time caused by cell phone use approximates that of a drunk driver. Prudence would indicate that far more resources should be devoted to public education, or enforceable laws to decrease the use of cell phones on the highway than to regulating the questionable risk of cell-phone induced cancers.

Analysis based solely on accident statistics, however, does not take into consideration emotional reactions to the very different nature of the two risks, the risk of immediate death or disability versus the risk of long-term debilitating or even deadly heath effects. In addition to cell phones, many other new or emerging technologies cause public concern, particularly exposure to chemicals that may have long-term carcinogenic, neurological or other harmful effects. These effects are studied in essentially the same manner as those of cell phones.

Accidental deaths and injuries inflicted immediately do not leave those who escape wondering whether there will be future consequences. Exposure to environmental contamination has the opposite affect, leaving those exposed to wonder if they will suffer long-term harm. An explosion or fire at a chemical plant, a smash-up of a freight train spewing chemicals from its tank cars, or the release of radiation to the atmosphere brings public reaction that is quite different from that of an automobile accident or an airplane crash.

Following the release of a contaminant, it may not be at all clear who has been harmed and who has escaped injury. The immediate loss of life may be small, or there may be none at all. Death and acute injury often are confined to the workers at the plant, railroad, or power station where the accident took place, or at worst to those living within close proximity. Likewise, physical destruction—if there is any—may be confined largely to property at the industrial complex where it occurred. But it is not the immediate harm that frightens large numbers of people; it is the concern that delayed health effects will develop, that in time what they have inhaled or ingested will cause debilitating disease or cancer.

The psychological impact of accidental contaminant release is likely to increase as time passes. As the cloud spreads or the leakage continues, the number of people who share the fear of exposure grows. Continuing media attention—alarmist or not—will tend to heighten anxiety. Public officials may find themselves in a quandary; often the seriousness of the release is at best unknown. Assessing the level of exposure may be difficult to immediately determine, and relating health effects to that undetermined exposure level is fraught with uncertainty.

Under such circumstances, prudence may dictate that the officials execute evacuation plans. Such uprooting, unfortunately, brings its own dangers—traffic accidents, falls, heart attacks, and looting—that may outweigh the uncertain toxicological effects of the contaminants. Moreover, even though the motivation for evacuation may be to assure that public health is being protected to the maximum, the media and the public are likely to come to very different conclusions: the situation must be truly life-threatening, or the evacuation would not have been ordered.

The problems faced in dealing with toxicological risk extend beyond accidents and their aftermath. Chronic exposure to dangerous substances in the workplace may also present health hazards, including the risk of severe occupational hazards to workers exposed to them every day. Some such risks are age-old and were long tolerated before scientists understood their dangers and the risks were brought under control. Health risks that arise from newer technologies—synthetic chemicals for example—are yet more difficult to quantify, as is knowing whether such effects even exist. With newly emerging technologies, there is no historical record of exposures to examine. Hence no opportunity makes it possible to examine health effects in a previously exposed population.

Even if scientists can estimate the level of exposure that has occurred, and if exhaustive studies provide a high level of confidence that the exposure was small enough to rule out significant heath effects, public officials are reluctant to say that those exposed suffered no risk at all. They may be justified only in stating that the chance of contacting cancer as a result of the contaminant is very small, or that the probability is much less than the normal incidence of the disease. The public, however, is not likely to be assured by statements that the added chance of cancer is less than one in ten thousand, or even one in a million. Even when based on broad scientific consensus, such statements are likely to evoke skepticism.

If clusters of cancers then do appear, the public opinion may attribute them to toxic contamination even though expert analysis may show that they are more reasonably attributed to chance fluctuations in the normal incidence of the disease. Such arguments based on probabilities may not be understood well and thus do nothing to reduce the lingering fears compounding the impact of the exposure. Clearly, the potential for delayed health effects presents technical and political issues not raised by incidents such as crashes, falls, drowning, and fires, for which the consequences are more immediately obvious.

* * * * *

Poisoning from materials found in nature, such as lead and mercury, has been a danger for millennia. With industrialization, however, exposure to these poisons increased, from mercury contained in coal smoke, for example, as well as from the increasing use of lead, not only in piping but also as additives to improve the quality of everything from paint to gasoline. But as living standards have risen, and other risks diminished, public pressure has grown to hasten the elimination of these toxic materials from the environment.

When the smoother operation of high-compression automobile engines resulting from lead additives in gasoline no longer justified the damage they caused to air quality, the additives were banned. Likewise, paint no longer contains lead, and plumbers no longer install lead piping. Only where it serves an essential purpose, and where there is no viable substitute, is lead allowed. Where lead remains in use, strict controls limit the occupational exposure to workers, minimize exposure of the public, and regulate disposal as well. However, such regulations have not removed the dangers from the lead in pipes and paint used prior to the ban in buildings, condos, and single-family dwellings. Removing the preexisting pipes and paint from those buildings was not required.

Campaigns have also reduced exposure to mercury. Use of once-ubiquitous mercury thermometers has been discouraged and banned. With accurate and inexpensive electronic thermometers readily available, the risk of mercury exposure from broken mercury thermometers is no longer acceptable. However, the mercury emitted in coal smoke remains an even more pervasive problem, one to which we shall return.

Asbestos is an example of a product in which adverse consequences can arise when a new technology is introduced with inadequate precaution and ignorance of its side effects. This fibrous natural material is strong, doesn't corrode, and has excellent properties for fire-proofing, heat insulation, and for strengthening concrete, tile, and other materials. Asbestos found widespread use in shipbuilding and other military applications during World War II, and its use skyrocketed in the civilian economy thereafter. It was used in everything from strengthening sewer piping and insulating heating and hot water pipes to fireproofing buildings and machinery.

Asbestos, however, is a powerful carcinogen if the fibrous dust generated in the manufacture, cutting, and installation of asbestos is inhaled. Recognition of this danger was delayed, however, because the resulting lung cancer didn't appear until twenty or more years after exposure. The cancer is no different than that caused by smoking. In fact, studies showed that smoking and asbestos inhalation have synergistic effects in increasing cancer rates.

Gradually evidence accumulated, pointing to the occupational hazard that the airborne fibers presented to factory workers exposed to asbestos dust on a daily basis. Regulatory actions banned this widely used insulation from building construction and other venues where a threat of public exposure might be significant. Stiff standards now regulate occupational exposure in those industries where asbestos remains in use because its benefits are substantial and results in no measurable public exposure.

As with other hazards, fundamental questions of safety must be addressed: Is banning asbestos use in new products and processes adequate, or should it be removed where it is already in place? Should it be removed from where it has already been installed as insulation or a fire retardant in schools, hospitals, offices, and apartments? Many school boards have wrestled with this issue. Is the danger to the students large enough to justify the large expense of eliminating the substance from the walls and ducts, an expense that may be so great as to cripple funding for other worthy projects to enrich education, and perhaps even those to improve the nutrition or address health issues of the students? If removal is required, can it be delayed or carried out over a period of years, or is the threat so ominous that immediate action is required?

Since asbestos had been so widely used in a variety of venues, the more difficult issue became to what extent and under what circumstances it must be removed. Here there is ongoing controversy. On the one hand, the thought of schoolchildren, hospital patients and others spending many hours in environments where asbestos is present in the walls and floor tile is disturbing. But subsequent study of the problem tends to conclude that the threat is predominantly from asbestos dust, an occupational hazard for those who work in manufacturing plants, and also for those present when asbestos-containing materials are cut, ground, or otherwise handled during removal from schools, hospitals, or houses.

Thus public perceptions and reality sometimes conflict. Even though we may fear the presence of asbestos, the reality is that leaving it in place in most cases causes less danger than the exposure during its removal. Containment by sealing it from direct contact from any abrasive action became an acceptable alternative in many buildings. The asbestos floor and ceiling tiles were covered by non-asbestos-containing tiles or sheeting materials to prevent the abrasive action of walking or general deterioration—such as cracking or breaking under moving furniture or heavy objects—to release fibers into the air. Asbestos-encased pipes were further covered with non-asbestos materials to prevent flaking as it deteriorated from abuse or aging. Loose asbestos that had been used as insulation in attics, walls, and ceilings was sealed with plastic foams that encased the fibers. These methods provided a less expensive alternative to removal.

* * * * *

Unlike recently introduced substances for which health effects are not yet fully known, the dangers of coal smoke have been with us since early in the Industrial Revolution. For nearly two centuries, the residents of cities and towns of any size have instinctively known that breathing polluted air was not good for their health. But it was accepted as an unfortunate byproduct of progress. Early in the Industrial Revolution, the economic progress of a community was judged by the amount of smoke billowing forth from its factory stacks.

Until recently, it was only when the health effect reached crisis proportions that decisive action was taken to decrease the danger. Such was the case, for example, in London in 1952, when a killer smog descended upon the city, causing widespread respiratory problems and taking the lives of as many as 12,000 residents. As a result of the crisis, the following years brought regulation to greatly reduce the amount of soft sulfur-rich coal used for home heating and to otherwise improve the city's air quality.

Coal smoke pollution is continuing to cause devastating health problems. In some parts of China, studies estimate that air pollution is cutting life expectancy by more than five years. In the industrialized West, progress in decreasing smoke has been made over the last century. Still, attempts to further reduce the harm directly attributable to pollutants contained in coal smoke meet with strong resistance, for economic well-being depends in part on the cheap electricity produced by burning coal.

Installation of effective pollution control, some would argue, would raise the price of electricity to unaffordable levels, and shifting rapidly to alternate fuels or means of production would create an industrial upheaval that would devastate the well-being of communities where coal is mined, shipped or burned. The conundrum is that, while the economic costs are immediate and obvious, the cost from illness, reduced lifespan and death are diffuse and difficult to measure in ways that focus public attention on the hazardous pollutants.

If coal had no impurities and combustion were perfectly safe, the only combustion product would be carbon dioxide (CO_2). Scientists agree that the presence of CO_2 in the atmosphere has no direct effects on human health. Rather, it is the accumulation of massive amounts in the atmosphere that is leading to global warming, arguably creating the greatest challenge to the well-being of mankind in the decades to come. Here, our concerns are with more immediate impacts; they deal with the effects of the other contents of coal smoke on human health.

Coal smoke contains a number of pollutants harmful to human health. Acidic vapors and very fine particulate matter—tiny particles much smaller than the width of a human hair—contain mercury, arsenic, lead, cadmium, and a host of other metals, many of them known to cause cancer. Up to half of the human exposure to mercury comes from burning coal; it is the most injurious of the pollutants, affecting the nervous system and estimated to cause cognitive difficulties for nearly 50,000 babies in the US each year.

Shortened lifespan, asthma attacks, and cancer have been linked to flue gas emissions from coal-burning power plants, as have respiratory, reproductive, and developmental problems. One study indicated that a single large coal-powered plant causes 7 premature deaths, 100 emergency room visits and more than 500 asthma attacks each year and results in annual costs of $42 million. A nationwide study concluded that the emission of particulate matter in the US results in nearly four billion dollars in public health damage each year. The variety and severity of the effects are illustrated by the Environmental Protection Agency's Heath Effects Pyramid shown in Figure 12.1.

Power plants vary widely in the ratio of pollutants emitted to electricity produced. In some coal-fired plants, substantial reductions in pollution have been achieved by installing advanced emissions control technology, such as equipment for flue gas scrubbing. But too often such pollution control measures are taken only after they are forced on companies by successful lawsuits. Despite some successful lawsuits, many older plants have no such controls. Only now is federal regulation of coal plant emissions coming into effect. The EPA's promulgation of regulations has been a long and politically charged process pitting citizen advocacy groups against the economic interests of electric utilities and coal producing states. Attempts by government

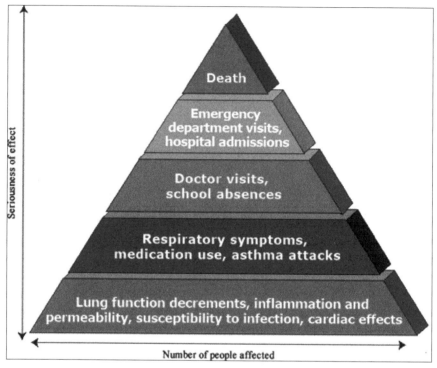

Figure12.1: Air Pollution Heath Effects Pyramid, *Environmental Protection Agency*

agencies to reduce coal smoke and other pollutants with well-known health effects often are challenged in the court and become subject of years-long litigation.

The delays encountered in reducing public exposure to dangerous pollutants is exemplified by the battle to reduce the hazards of coal smoke. In 1990, more than 20 years ago, Congress amended the Clean Air Act, mandating that the EPA reduce mercury and toxic air pollution. Hearings, lawsuits, and the need for extensive cost-benefit analysis caused repeated delays in the rulemaking process needed to implement the Congressional mandate. Finally, in response to a lawsuit by the Sierra Club and backed by other advocacy groups, the District Court of the District of Columbia set a deadline of the end of 2011 for the EPA to finalize the rules.

Those rules require all coal-burning power plants to install the best available methods for limiting flue gas emissions of mercury and other toxic materials. Nationwide, installation of the equipment should reduce mercury and acid gas emissions by approximately 90 percent. The EPA analysis predicted that as a result there should be a reduction of premature deaths by more than 10,000 per year, of asthma attacks by 130,000 per year, and of heart attack by nearly 5,000 per year. The EPA's cost benefit analysis concluded that the controls would result in health benefits of between 22 and 54 billion dollars, while the cost of the controls would be an estimated 1.4 billion dollars. Thus the ratio of benefits to cost would be between five and thirteen to one! The largest impact would derive from the prevention of premature death from fine particle pollution.

Even with irrefutable results in hand, implementation of the rules has been delayed for more than a year by lingering political argument that the adverse impacts on jobs and on the economies of coal-producing and -burning areas of the country would be too large. The occupational risks of coal mining, ranging from accidents to black lung disease, in those same localities were not taken into consideration.

* * * * *

None of the preceding examples involve immediate fatalities as do the daily auto accidents, drowning, falls, and fires that account for the bulk of the accidental death toll each year. Nor are they frequently related to spectacular disasters leading to immediately visible damage and loss of life. The exceptions are those fires and explosions from which delayed effects may arise from the release of toxic chemicals to the environment. For the most part, the losses of life are more insidious, occurring over extended periods and leaving many with the worry that their health has been put in jeopardy.

Worries over the health hazards that technology may cause are many and varied. They range from concern over newer technology, such as cell phones—even where exhaustive research has found no evidence that a health effect exists—to the fear of radiation exposure, which far exceeds the dangers indicated by scientific analysis.

Lead, mercury, asbestos, fine particulate matter and chronic exposure to other more common environmental contaminants pose far more pervasive health problems, particularly for those whose work or the location of their homes subjects them to high concentrations. Steady if uneven progress has been made in reducing exposure from these contaminants.

Arguably the greater danger comes from the pollutants closest to home, to which little attention is paid and little recognition given to the hazards they present. Yet it is the hazards to which the public has grown most accustomed that in fact take a larger toll in disability and lost years of life. Concentrations of the indoor pollutants found in homes, where particularly women and vulnerable young children spend most of their time, are a worldwide hazard.

In less-developed countries, indoor air pollution is at its worst; often it is worse than what is found outside. Low-temperature combustion of solid materials—wood, coal, dung, or crop waste for heat—create copious amounts of fine particles that penetrate and damage the lungs. Furthermore, cooking in poorly ventilated areas produces a great deal of acrolein, an acrid-smelling chemical that causes respiratory and heart problems. Likewise, the fumes from poorly controlled kerosene lighting lead to dire health effects from breathing the air.

World Health Organization studies document that 3.5 million premature deaths result each year from the poor air quality inside structures inhabited by most people. In fact, an argument can be made that much of the added life expectancy in the developed world has resulted from central heating, the conversion from wood and coal to gas and oil, from candles and kerosene lamps to electric lighting, and other improvements that have reduced the level of indoor pollutants. Nevertheless, risks remain, even in highly developed nations.

Indoor air pollution is a little-publicized cause of premature death and disability, while we focus our concern on what we encounter when we step outside our door. The Environmental Protection Agency, however, reports that the air inside the average American home is up to five times as polluted as the air outside. Extensive research at the EPA and other organizations documents the scope of these health hazards. Even where tobacco smoke, the most dangerous of contaminants, has been eliminated, everyday activities produce hazardous pollutants.

Cooking ranks at the top of the list in producing indoor pollutants. Whether in frying, broiling, or toasting, daily food preparation sets fats and oils and even carbohydrates afire, producing acrolein as well as fine particulate matter that penetrates the lungs and damages the respiratory and vascular systems. Air fresheners, cleaning fluids and other household materials add to the burden of indoor air pollution. Formaldehyde, emitted from carpets, upholstery, fiberboard furniture, and paneling, compounds the toxic mix.

In addition to creating nitrous oxides and carbon monoxide, open flames, such as from a gas stove or water heater, compounds the hazards further, yielding an array of yet more toxic gases as fumes from formaldehyde, cleaning materials, paint, glue, and other household materials pass through the open flame. Some estimates put the toll of premature death and disability in the US and other industrialized countries at the same level as the lives lost in motor vehicle accidents.

We can take precautions to reduce the presence of hazardous pollutions in our homes. Ensuring that combustion in gas appliances, furnaces and fireplaces is properly controlled and that the gases are properly vented should be placed high on the list for improvement. Adequate ventilation, particularly when using strong cleaning materials, paint, glue, or other fume-producing products, is also important. But here tradeoffs again come into play; as we strive to make our home more energy-efficient, we are likely to reduce ventilation, hence bottling up the indoor pollutants.

Beyond forbidding smoking, the most important preventive measure in reducing exposure to pollution in the home is to have an effective exhaust system over the stove, and to use it routinely. Yet this precaution, which may be as important in reducing premature death as wearing seat belts when we drive, is nearly universally ignored.

With health hazards and accidents alike, our worries tend to concentrate on spectacular, poorly understood, or bizarre risks to which the media gives most attention. They are the risks that we feel powerless to control and need stronger government regulation to reduce to manageable proportions. Yet, we tend to ignore the dangers we can control through our personal behavior and vigilance. Thus, just as our poor driving habits account for more deaths each year than airliner crashes or train wrecks, our lack of attention to controlling widespread pollutants at home claims more lives than chemical spills or the exotic environmental contaminants that receive the bulk of headline news.

CHAPTER 13

Toward a Safer Society

WE RETURN ONCE again to the question posed by the title of this book: how safe is safe enough? It is actually a two-part question: how safe is technology, and is that safe enough? The first is difficult to answer, particularly for new or rapidly changing technologies. But eventually the risks of new technologies are identified, statistics gathered and estimates made of the lives lost, disabilities incurred and health hazards inflicted. How we view whether the technology in question is safe enough depends on a host of issues such as who we are, our deep-seated attitudes toward risk, our economic interests, our views toward privacy, liberty, regulation and our belief regarding the right of individual choice and whether that right supersedes the safety of society at large. And always, there are tradeoffs between our willingness to tolerate the risks, regulations, and the many needs and desires that motivate much of what we think and do.

Our view of whether a particular technology is safe enough depends crucially on its nature. The more spectacular the accident, the larger the loss of life, the more bizarre the circumstance, the more media attention it will undoubtedly receive. And it is our nature to overweigh the importance or the risks of such rare events. The same can be said for new and mysterious risks about which little is known, particularly those that appear to have no immediate effect but pose the possibility of dire health effects long into the future. Such things are frightening, and more often than not they are out of our personal control. Thus we may be inclined to demand action through stricter laws, more enforceable regulation or other means for which we can lobby and demand political action.

Conversely, we tend to take in stride risks that are common, that we face on a daily basis. When such accidents occur, rarely do they take more than one or two lives at a time. Unless they involve a small child or have some other poignant aspect, they receive little or no media attention. Unless they befall our friends or family, we give them only passing attention. Moreover, reducing our personal risk from these common mishaps would require our being more vigilant, more cautious or otherwise modifying our own behavior, something we are not prone to do. Thus, as statistics show, motor vehicle accidents, household falls and fires, accidental poisoning, and drowning account for far more deaths than spectacular disasters or bizarre events. Similarly, more years of disease and shorter lifespans result from our chronic exposure to commonplace pollutants—particularly those in our homes—than to massive chemical spills or to the insidious affects of exotic albeit narrowly employed toxins.

No doubt, the greater tolerance for commonplace risks and mundane accidents stems from the belief that we are in control. We feel that we can be careful and minimize the risks if we choose to; we could drive more carefully, be more careful in avoiding falls, be more conscientious in handling and storing flammable or potentially poisonous materials, even if in fact we are not. Likewise, if we take chances jet-skiing, snowmobiling, hunting, or other risky sporting activities, we tolerate the risks because they are voluntary, an integral part of the thrills that we seek.

But if a chemical plant explodes, a ferry boat sinks or a bridge collapses, that is much more frightening, for those killed or injured had no control over the risk. And even in the most common of accidents—motor vehicle collisions—we are less fearful when we hear of one caused by driver error rather than vehicle failure, because we can convince ourselves that we would have taken precautions to avoid it. But if brake failure or a steering malfunction was the cause, we can easily imagine ourselves caught in a similar situation in which our fate is out of our control.

We feel more sympathy for those who are innocent. We are horrified when airline passengers or chemical plant employees die in disastrous circumstances. But at least they had subjected themselves to the risk to gain some benefit. Our horror is yet stronger when innocent bystanders are killed, those not benefiting directly from the technology—such as those on the ground killed by a crashing aircraft or living near a railroad when a tank car spews deadly fumes into the atmosphere. They were simply unfortunate enough to be in the wrong place at the wrong time, and likely were ignorant of the risk's existence. Likewise, we react strongly to risks imposed on children, the elderly, the ill, and others not capable of looking after themselves.

Public officials are limited in what they can do to reduce the incidence of accidents caused by attitudes and behavior rather than equipment failure. Many officials, for example, may favor random stops for alcohol testing, stricter enforcement of traffic laws, and greater penalties for infractions to reduce the number of automobiles fatalities. But beyond a certain point, resistance grows as motorists lobby against the inconvenience and limitation placed on their freedom by such stricter laws. Political resistance increases and often expresses itself as budget limitations placed on regulatory agencies as the public increasingly views implementation of the programs as coercive rather than encouraging safer behavior. We value our privacy and freedoms; this also limits what we allow government to do in the name of reducing the accidental risks that we face.

How demanding the calls for stricter regulation are depends on who is affected as well as on the nature of the risk. The search for cancer induced by cell phones receives plenty of funding without evidence of harm because virtually everyone in the more affluent strata of society relies heavily on cell phone use. But we resist the limitations on our behavior that banning their use while driving would bring, even as many hundreds die each year from cell phone use in automobiles.

Public and corporate officials, entertainers and prominent citizens in many walks of life fly frequently, assuring that strict regulation will continue to ensure the safety of commercial aviation. Interstate bus services receive less scrutiny because generally their riders are poorer and have little political clout. Likewise, occupational risks and the agencies that attempt to reduce them receive scant political attention. The specific risk may be great, but if it affects only a small fraction of the nation's workers—and they have little political influence relative to those who employ them—the path toward reducing the risk is plagued by difficulties.

Benefits and financial costs result from any technology, and actions to enhance safety change the ratio of the two. Using formal constructs such as statistical life expectancy, governmental agencies try to make formal assessments to determine whether the lives saved and injuries avoided by moving to stricter safety regulation will outweigh the financial costs that will be incurred. But most often those costs and benefits are calculated for society as a whole. And the conclusions of formal analysis are confounded politically because the safety benefits often fall on different segments of society from those who bear the costs. Who enjoys the benefits of technology versus who is subjected to its risks is critically important to the political processes by which we as a society determine how safe we require technology to be.

Legislators from states where coal is extracted will react differently to proposals for more stringent mine safety regulation than those from localities where it is transported and burned. A miner whose livelihood depends on the demand for coal will have views that differ sharply from those of an urban dweller subjected to the air contamination caused by a nearby coal-fueled power plant.

Corporate executives resist safety regulations that they feel may harm their competitive position, and cost concerns are likely to dampen calls for more stringent regulation by those who buy the products. Similarly, local fire marshals' views on the enforcement for smoke detector rules are likely to be quite different from those of some homeowners who believe strongly that their right to personal choice must be protected. Skydivers, bungee jumpers and operators of recreational off-road vehicles object to safety restrictions that they feel degrade the full thrill they seek. And the list goes on . . .

Even in our own decisions, we must make tradeoffs between safety and other considerations. When buying an automobile, we of course would like to pick the safest model possible, but we must weigh safety against a price that may be unaffordable, as well as a desire for fuel efficiency, whether the desire is motivated by reducing our outlay for gasoline or by contributing less to global warming. And on the road as the weather turns bad, we must decide how much risk to accept in order not to arrive late at a function that is important, economically or socially. Frequent blazing wood fires in an open fireplace are esthetically appealing experiences, and we likely don't want to diminish our pleasure by focusing on the invisibly tiny particulate matter that may be engulfing the room and increasing our susceptibility to respiratory illness.

Technologists learn from their mistakes. Whether accidents are large or small, whether they are frequently occurring common causes of death or rare but deadly catastrophes, methods to make technology safer emerge from the study of what went wrong. The question then becomes how to implement the newfound knowledge. Requiring new machines or structures that haven't yet been designed or built to adhere to this newfound safety knowledge is the least costly. Retrofitting technology that is presently in use is more difficult and expensive. And a related question greatly affects the economics of implementation: How soon must retrofitting be complete? Most difficult of all is facing the decision of whether structures or machines are deemed too expensive or impossible to upgrade to meet the more stringent safety criteria, so that they must be scrapped.

These three degrees of implementation must be considered in terms of costs and benefits, and the details vary greatly depending on the nature of the problem. Consider two widely differing technologies. Whether a swimming pool drain or a baby crib, when the Consumer Product Safety Commission rules that it presents an unacceptable risk, the agency can ban production unless acceptable modifications are made in the design. More strongly and with increasing frequency, it can require recall of the products already in use for modification or replacement. For many consumer products, replacement is much less expensive than retrofitting. The difficulty is instead in locating the owners of the defective product and motivating them to replace it. Finally, in some cases—as in drop-side baby cribs—the agency may ban the product altogether.

In contrast, consider the threat of natural disasters, such as tornadoes. In high-risk localities, should shelters be required only in new home construction, or should older homes be required to be retrofitted? If retrofitting is required, how many years should be allowed for it to be completed? As an alternative, should retrofitting be required only when the house is sold to a new owner? If it is exorbitantly expensive to build a shelter as an integral part of some housing, such as house trailers, what should be required to provide protection to those residents? Whether for consumer products or automobiles, or for protection from tornadoes or earthquakes, implementing tightened safety standards faces many and varied challenges.

* * * * *

Even those people whose careers have immersed them in understanding technology and those steeped in the quantitative methods of risk assessment can react subconsciously to signs of dangers inherent in technology, and the author is no exception. For risk assessors, engineers and others in similar fields, the risks are all the more ubiquitous.

Just as physicians are prone to hypochondria, engineers live by the adage that if anything can go wrong, it will. Noticing the onset of a strange rattle while driving at expressway speeds or while waiting on the tarmac as the pilot announces a delay while the ground crew attends to a mechanical problem may trigger worry. When I pick up my car following maintenance, I wonder if the brakes have been properly reassembled following lining replacement, and the bolts properly tightened following tire rotation. When I go in for a medical procedure—an X-ray or CAT scan—I think about the deaths that have resulted from poorly designed equipment that lead operators to make understandable but fatal mistakes.

When I board an aircraft, I observe the rows of rivets along the fuselage and think of the Aloha airliner becoming "unzipped" and wonder about the underlying state of corrosion not visible from the exterior of the plane. As I pass the cockpit door, I wonder whether the pilot's skills have waned from overreliance on automated systems; as the steward readies the galley for takeoff, I think of the case studies of fires caused by faulty circuitry and personnel errors. I look out the window at the engine and think of bird strikes and the geese that disabled flight 1549.

Sometimes I envy fellow passengers, those not professionally attuned to thinking about potential failure modes everywhere they look. But I force myself into a more analytic frame of mind to focus on just how rarely the life-threatening failures on aircraft occur. It helps to be familiar with flying statistics, that every year many millions of people fly on US scheduled airline flights, and frequently a year or more will pass without a single crash. I think of the pilots and cabin crew and appreciate that they make several takeoffs and landings each day. Likely all of them will complete many years of this routine and go on to their retirement never having been aboard an aircraft that came close to crashing. Comforted by these thoughts, I am able to sit back and relax with the innate fear of flying that lurks within us all firmly in check.

At least we understand aircraft crashes, train wrecks, and similar disasters; even if we experience some degree of fear while flying, riding or participating in other activities where disaster is a possibility—however remote—we can relax when we are at home or otherwise not exposed to the perceived dangers. But even those professionally conditioned to probability and the quantitative evaluation of risk may experience a mild feeling of dread that other technological risks are present—those that have not yet been identified, that may come from new untested technology, or from technology already in use for some time, but that may result in delayed health effects that are not discovered until years after exposure have taken place.

I relate to such anxiety, for I have experienced it first hand: many years ago, in the mid 1970s, my wife was diagnosed with thyroid cancer. The cause was X-ray treatments that she had had for a very mild case of teenage acne. Radiation therapy was a common treatment for children and adolescents with acne, inflamed tonsils, or adenoids in the 1950s before the medical community became aware of the side effects from excessive exposure to such radiation. An epidemic of thyroid cancer ensued twenty years later. Fortunately, when treated early, thyroid cancer is rarely fatal, and most patients recovered fully.

The uncertainties in new technology can bring foreboding not only to the public at large, but also to those who have devoted their careers to studying its risks and benefits. Particularly scary are new substances—synthetic chemicals for example—the long-term side effects of which may not be understood for years into the future. And there may be cause for worry: in the past, new technologies have been introduced without adequate study, with dire results. But the fact that a compound is found in nature does not guarantee a lack of risk in use. Lead and asbestos are classic examples.

In the last half-century methods for dealing with the potential health risks of technology have improved as science has focused efforts to foresee future health effects of new medical procedures and of substances introduced into the environment. Nevertheless, the dread of a future cancer epidemic or infant birth defects lingers in our minds. The concern is understandable. Keep in mind, however, that there are real, understood technological risks that have a substantial impact on health and longevity. As a society, however, we are not doing more to ease the impact because of the large economic costs that such efforts are likely to bring.

The smog caused by large-scale burning of coal for power production and other uses, and the particulate matter coming from internal combustion engines offer pertinent examples. The adverse health effects of these are large and well-documented, leading to asthma attacks, emergency room visits, and documentable decreases in life expectancy from respiratory and cardiovascular disease. Our efforts might be better spent in working to reduce these known hazards than in ruminating about the potential risks from newer technologies for which—despite their efforts—scientists have yet to find evidence of harm.

While we must keep the potential for disaster in perspective, and the worries of poorly understood contaminants in tow, we must also be vigilant in our everyday activities, for it is there that the largest risks lie; these activities are generally under our control. We should install and use handrails; we must walk carefully on slippery surfaces; we should be diligent in replacing smoke detectors on a regular basis. When tired, distracted, or under the influence of alcohol, we must avoid activities where lack of attention, slowness of reflexes or clouded judgment could lead to an accident. If we are hunters, we must make sure that our rifles are stored unloaded in locked cabinets; if we are boaters, we must insist that lifejackets be worn and that deteriorating weather conditions be taken into account.

But for most of us, driving is where we must be most vigilant. Although one auto journey may have a very small likelihood of resulting in a serious accident, most of us take many thousands of trips over our lifetime over which the probability of a serious accident rises. We must take care each time we merge onto an expressway, check blind spots each time we change lanes, and avoid being distracted at these crucial moments by fiddling with the radio or, worse, talking on a cell phone.

Writing in *The New York Times*, Pulitzer Prize-winning author Jared Diamond summed up the situation well:

> . . . we exaggerate the risks of events that are beyond our control, that cause many deaths at once or that kill in spectacular ways—crazy gunmen, terrorists, plane crashes, nuclear radiation, genetically modified crops. At the same time, we underestimate the risks of events that we can control ("That would never happen to me—I'm careful") and of events that kill just one person in a mundane way.

Professor Diamond is speaking, of course, of those risks—many of which are discussed in this book—that may have an immediate impact, however improbable, on us individually. Other risks, which fall outside the scope of this volume, are longer term, and are existential threats to humankind as a whole: if the nations of the world stumble into global nuclear war, the devastation would be unimaginable. And if they do not unite to deal effectively with global warming, climatic change is likely to cause flooding of densely populated seacoasts, ravaging weather-induced natural disasters and widespread famine. While assuring that technology's immediate risks are maintained at acceptable levels, we must focus our efforts more strongly to deal with these longer-term threats if we are to bequeath a livable planet to future generations.

Acknowledgments

Many friends, colleagues and students have contributed to my effort to provide an evenhanded examination of the risks and benefits that accompany technological development. My years of collegial discussions with fellow faculty members at Northwestern University and the resources and staff of its library have been indispensible in bring this project to fruition. Interactions with the many students whom I've had the privilege to teach, particularly those taking the reliability engineering course—ME 359—that I've taught for many years, have done much to advance my views on numerous topics in this volume. I'm also indebted to my agent, Roger Williams, and to my editor, Niels Aaboe; both have done much to move this project from concept to completion.

For more than a year, Roger Blomquist, Nicholas and Marcy Collins, and Lloyd Evans have read earlier drafts and provided suggestions and criticisms that have made the final manuscript better than it would otherwise have been. Their efforts notwithstanding, I take full responsibility for any errors and shortcomings that remain. The book has also benefited from the historical perspective of my daughter Elizabeth Lewis Pardoe, and from the scientific input of my son Paul Lewis. But most of all, it has been Ann, my wife of more than 50 years, who has contributed to this effort. In addition to offering encouragement, she has read both the draft and the near-final manuscript in their entirety, making many suggestions, including eliminating passages that only engineers could understand, and making the text more accessible to the general reader.

Elmer E. Lewis
Evanston, Illinois, 2014

Selected Bibliography

CHAPTER 1

Armytage, Walter Harry Green. *A Social History of Engineering.* Cambridge, MA: MIT Press, 1961.

Basalla, George. *The Evolution of Technology.* Cambridge UK: Cambridge University Press, 1998.

Carling Tobias, and Robert Udelsman. "Thyroid tumors." *Cancer: Principles and Practice of Oncology.* 9th ed. V. T. DeVita Jr., T. S. Lawrence, and S. A. Rosenberg (eds.). Philadelphia: Lippincott, Williams & Wilkins, 2011.

Faith, Nichoias. *Derail: Why Trains Crash.* London: Macmillan, 2000.

Klemm, Friedrich. *A History of Western Technology.* Cambridge, MA: MIT Press, 1964.

Kranzberg, Melvin, and Caroll W. Pursell Jr. (eds.). Technology in Western Culture, Vols. 1 &2, Oxford, Oxford University Press, 1967.

Lewis, Elmer E. *Masterworks of Technology: The Story of Creative Engineering Architecture and Design.* Amherst New York: Prometheus Books, 2004.

Verhovek, Sam Howe. *Jet Age: The Comet, the 707 and the Race to Shrink the World.* New York: Penguin, 2010.

CHAPTER 2

Aircraft Accident Report: Aloha Airlines, Flight 243, Boeing 737-200, AAR 8903. Washington D.C.: National Transportation Safety Board, April 1988. http://www.ntsb.gov/investigations/summary/aar8903.html

Aviation Accident Report: Loss of Thrust in Both Engines After Encountering a Flock of Birds and Subsequent Ditching on the Hudson River: US Airways Flight 1549 Airbus A320-214 N106US Weehawken NJ. Washington D.C.: National Transportation Safety Board, May, 2010. http://www.ntsb.gov/doclib/2010/aar1005.pdf

Dean, Kenneson G., Lawrence Whiting, and Haitao Jiao. "An Aircraft Encounter with a Redoubt Ash Cloud." *Volcanic Ash and Aviation Safety: Proc. 1ˢᵗ Int. Conf.,* US Geological Survey Bulletin 2047, 1991.

Lewis, Elmer. E. *Introduction to Reliability Engineering,* 2ⁿᵈ Ed. New York: Wiley, 1994.

Marshall, R. D. *et. al., Investigation of the Kansas City Hyatt Regency Walkway Collapse.* Washington D.C.: The National Institute of Standards and Technology, May 1982. http://fire.nist.gov/bfrlpubs/build82/art002.html {keep on one line.

Massimo, Livi-Bacci. *A Concise History of World Population.* Translated by Carl Ipsen. Oxford: Blackwell, 1992.

Perrow, Charles. *Normal Accidents: Living with High-Risk Technologies.* Princeton: Princeton University Press, 1999.

Wittstein, Theodor. "The Mathematical Law of Mortality." *Journal of the Institute of Actuaries and Assurance Magazine,* London, 1883.

CHAPTER 3

2010 Annual Report of ATV-Related Deaths and Injuries. Consumer Product Safety Commission, December, 2011. www.spsc.gov/PageFiles/108609/atv2010.pdfBethesda

Greenhouse, Steven. "Pilot Error Is Blamed in Airbus Crash." *The New York Times,* June 28, 1988.

Hinch, John, *et. al. Air Bag Technology in Light Passenger Vehicles.* Washington D.C. National Highway Traffic Safety Administration, June 2001.

Price, Don K. *The Scientific Estate.* Cambridge, MA: Harvard University Press, 1965.

Schmitt, Vernon R., James W. Morris, and Gavin D. Jenney. *Fly-by-Wire: A Historical and Design Perspective.* Washington D.C.: Society of Automotive Engineers, 1998.

Smith, Merrit Roe (ed). *Military Enterprise and Technological Change: Perspectives on the American Experience*. Cambridge, MA: MIT Press, 1987.

Whittle, Richard. *The Dream Machine: The untold History of the Notorious V-22 Osprey*. New York: Simon & Schuster, 2010.

CHAPTER 4

Collapse of I-35W Highway Bridge Minneapolis Minnesota, August 1, 2007. Washington D.C.: National Transportation Safety Board Accident Report NTSB/HAR-08/03, 2008. www.dot.state.mn.us/35Swbridge/finalreport.pdf

FAA's NextGen Implementation Plan. Washington D.C.: Federal Aviation Administration. March 2013. www.faa.gov/nextgen/implementation/media/NextGen_.

Metal-on-Metal Hip Implant Systems, Federal Drug Administration Executive Summary Memorandum, Silver Spring, MD: Federal Drug Administration, 2012.

National Highway Traffic Safety Administration Toyota Unintended Acceleration Investigation: NASA Engineering and Safety Center Technical Assessment Report, January, 2011.

Petroski, Henry. *To Engineer Is Human: The Role of Failure and in Successful Design*. New York: St. Martin's Press, 1985.

Walton Mary. *Car*. New York: Norton, 1997.

CHAPTER 5

"Alberta woman dies from chemotherapy overdose," Canadian Broadcasting System News, Aug 31, 2006.

Bogdanich, Walt. "Radiation Offers New Cures, and Ways to Do Harm." *The New York Times*, January 23, 2010.

Final Report: On the accident on 1ˢᵗ June 2009 to the Airbus A330-203 Flight AF 447 Rio de Janeiro-Paris. Bureau d'Enquêtes et d'Analyses pour la sécurité de l'aviation civile, July 2012. http://www.bea.aero/en/enquetes/flight.af.447/rapport.final.en.php

Onishi, Norimtsu, Christopher Drew, Matthew L. Wald, and Sarah Maslin Nur. "Terror on Jet: Seeing Water, Not Runway." *The New York Times*, July 7, 2013.

Three Mile Island: A Report to the Commissioners and to the Public. Washington D.C. US Nuclear Regulatory Commission, 1980. www.threemileisland.org/downloads/354.pdf

To Err Is Human: Building a Safer Health System. Washington D.C.: The National Academy Press, 2000.

Whittingham, Robert B. *The Blame Machine: Why Human Error Causes Accidents.* Burlington, MA: Elsevier Butterworth-Heinemann, 2004.

CHAPTER 6

Bennett, Burton, Michael Repocholi, and Zhanat Carr (Eds.). *Health effects of the Chernobyl Accident and Special Health Care Programs.* Geneva: World Health Organization, 2006.

Broughton, Edward. "The Bhopal disaster and Its Aftermath: a Review." *Environmental Health,* 2005. http://www.ncbi.nlm.nih.gov/pmc/articles/PMC1142333/

Coronel, Sheila S. "Searchers Find No Trace of 1,500 From 2 Ships Sunk in Philippines." *The New York Times,* December 22, 1987.

Deep Water: The Gulf Oil Disaster and the Future of Offshore Drilling. Washington D.C.: US Government, National Commission on the BP Deepwater horizon oil Spill and Offshore Drilling, Progressive Management, 2011. http://www.gpo.gov/fdsys/pkg/GPO-OILCOMMISSION/content-detail.html

Lustgarten, Abraham. *Run to Failure: BP and the Making of the Deepwater Horizon Disaster.* New York: Norton, 2012.

Van Heerden, Ivor. "The Failure of the New Orleans Levee System Following Hurricane Katrina and the Pathway Forward," *Public Administration Review 67,* December, 2007.

Yardly, Jim. "Report on Deadly Factory Collapse in Bangladesh Finds Widespread Blame." *The New York Times,* May 22, 2013.

CHAPTER 7

Burke, John G. "Bursting Boilers and the Federal Power." *Technology and Culture,* Vol. 7 No. 1 , 1966.

Cheit, Ross E. *Setting Safety Standards: Regulation in the Public and Private Sectors.* Berkeley: University of California Press, 1990.

Slovic, Paul. *The Perception of Risk.* London: Earthscan, 2000.

Sunstein, Cass R. *Risk and Reason: Safety, Law, and the Environment.* New York: Cambridge University Press, 2011.

Valuing Mortality Risk Reductions for Environmental Policy: A White Paper. Washington D.C.: US Environmental Protection Agency Center for Environmental Economics, 2010.

Wilson, Richard, and Edmund A. C. Crouch, *Risk–Benefit Analysis.* Cambridge MA: Harvard University Press, 2001.

CHAPTER 8

"4 Killed in Crash, Not Wearing Seat Belts." KUSA-TV Denver, July 7, 2013.

Bradsher, Keith. *High and Mighty: the Dangerous Rise of the SUV.* New York: Public Affairs, 2002.

Evans, Leonard. *Traffic Safety.* Bloomfield Hills, MI: Science Serving Society, 2004.

Hedlund, James. "Review: Risky business: safety regulations, risk compensation, and individual behavior." *Injury Prevention,* 2000.

Lives Saved Calculations for Seat Belts and Frontal Air Bags. Washington D.C.: National Highway Traffic Safety Administration, DOT HS 811 206, December 2009. www-nrd.nhtsa.dot.gov/Pubs/811206.pdf

Shinar, David. *Traffic Safety and Human Behavior.* Bingley, UK: Emerald Group Publishing, 2007.

Traffic Safety Facts 2011: A Compilation of Motor Vehicle Crash Data from the Fatality Analysis Reporting System and the General estimates System. National Highway Traffic Safety Administration, October, 2013. www-nrd.nhtsa.dot.gov/Pubs/811659.pdf

CHAPTER 9

Berkes, Howard, and Jim Morris, "Fines Slashed in Grain Bin Entrapment Deaths," *National Public Radio Special Series, Buried in Grain.* Washington D.C.: National Public Radio, March 24, 2013. www.npr.org/series/174755100/buried-in-grain

Cooper, Michael, Gardiner Harris and Eric Lipton. "In Mine Safety, a Meek Watchdog." *The New York Times,* April 10, 2010.

Hughes, Jennifer V. "New Laws Try to Keep Swimmers Safe." *The New York Times,* June 15, 2008.

OSHA Fact Book. Occupational Safety and Health Administration, December 2008.

Schroeder, Tom. *Consumer Product-Related Injuries and Deaths in the United States: Estimated Injuries Occurring in 2010 and Estimated Deaths Occurring in 2008.* Washington D.C.: US Consumer Product Safety Commission, February, 2012. www.osha.gov/as/opa/OSHAfact-book-stohler.pdf

Urbina, Ian. "As OSHA Emphasizes Safety, Long-Term Health Risks Fester." *The New York Times*, March 30, 2013.

CHAPTER 10

Aircraft accident Report: Pan American World Airways Boeing 747 N737PA; KLM Royal Dutch Airlines Boeing 747 PH-BUF; Tenerife, Canary Islands, March 27, 1977. Washington D.C.: Air Line Pilots Association Engineering and Air Safety, 1978.

Austin, Ian. "Deaths Climb from Disaster in Quebec." *The New York Times*, July 10, 2013.

Best, Richard. *Investigation Report on the MGM Grand Hotel Fire.* Quincy, MA: National Fire Protection Association, November, 1980.

Bibel, George. *Trainwreck: the Forensics of Rail Disasters.* Baltimore: Johns Hopkins University Press, 2012.

Chiles, James R. *Inviting Disaster: Lessons From The Edge of Technology.* New York: Harper Business, 2001.

Gibney, Frank (ed). *Catastrophe! When Man Loses Control.* New York: Bantam/Britannica Books, 1979.

Mouawad, Jad. "Too Big to Sail? Cruise Ships Face Scrutiny." *The New York Times*, October 27, 2013.

MV Herald of Free Enterprise: Formal Investigation, Report of Court No. 8074. London: Department of Transport, Her Majesty's Stationery Office, 1987.

Slovic, Paul. *The Perception of Risk*, London: Earthscan, 2000.

Smothers, Ronald. "Dozens Are Killed in Wreck of Train in Alabama Bayou." *The New York Times*, September 23, 1993.

CHAPTER 11

Allen, Richard. "Seismic Hazards: Seconds Count." *Nature*, October 2, 2013.

Fradkin, Philip L. *The Great Earthquake and Firestorms of 1906: How San Francisco Nearly Destroyed Itself.* Berkley: University of California Press, 2005.

Green, Nathan C. (ed.). *Story of the 1900 Galveston Hurricane*. Gretna: Penguin, 2000.

Hurricane Basics. National Oceanic and Atmospheric Administration, 1999.

Hurricanes, Heat Waves, Snowstorms, Tsunamis and Other Natural Disasters. New York: Palgrave and Macmillan, 2012.

Magourney, Adam. "The Troubles of Building Where Faults Collide." *The New York Times*, November 30, 2013.

Penna, Anthony N. and Jennifer S. Rivers. *Natural Disasters in a Global Environment*. New York: Wiley, 2013.

Preliminary dose estimation from the nuclear accident after the 2011 Great East Japan earthquake and tsunami. Geneva: World Health Organization, 2012.

CHAPTER 12

Alleman, James E., and Brook T. Mossman. "Asbestos Revisited." *Scientific American*, 277: 54—57. 1997.

Clarkson, Thomas W., Lars Friberg, Gunnar F. Nordberg and Polly R. Gager (eds.). *Biological Monitoring of Toxic Metals*. New York: Plenum Press, 1988.

Emissions of Hazardous Air Pollutants from Coal-Fired Power Plants. Needham, MA: Environmental Health & Engineering 2011.

Lomborg, Bjorn. "The Poor Need Cheap Fossil Fuels." The *New York Times*, December 3, 2013.

Schüz, Joachim, Rune Jacobsen, Jorgen H. Olsen, J. D. Boice, J. K. McLaughlin, and Christoffer Johansen. "Cellular Telephone Use and Cancer Risk: Update of a Nationwide Danish Cohort." *Journal of the National Cancer Institute* 98 (23), 2006.

Scoullos, Michael, Gerrit H. Vonkeman, I. Thornton, and Z. Makuch. *Mercury — Cadmium — Lead Handbook for Sustainable Heavy Metals Policy and Regulation*. New York: Springer, 2002.

Smith, Peter Andrey. "The Kitchen as a Pollution Hazard," *The New York Times*, July 22, 2013.

CHAPTER 13

Diamond, Jared. "That Daily Shower Can Be a Killer." *The New York Times*, January 28, 2013.

Hedlund, James, "Risky Business: Safety Regulations, Risk Compensation, and Individual Behavior." *Injury Prevention* 6, 2000.

Jones-Lee, Michael W. *The Economics of Safety and Physical Risk*. Oxford: Basil Blackwell, 1989.

Kahneman, Daniel. *Thinking, Fast and Slow*. New York: Farrar, Straus and Giroux, 2011.

Ropeik, David. *How Risky Is It, Really?: Why Our Fears Don't Always Match the Facts*. New York: McGraw-Hill, 2010.

Index

G